Genes Unveiled: The Power of Personalized Medicine in Your Hands: Unlocking the Secrets of Your Unique DNA for Optimal Health and Well-being

Kari D. Smith

COPYRIGHT

All rights reserved. No part of this publication may be reproduced, distributed, or transmitted in any form or by any means, including photocopying, recording, or other electronic or mechanical methods, without the prior written permission of the publisher, except in the case of brief quotations embodies in critical reviews and certain other noncommercial uses permitted by copyright law.

Copyright © (Kari D. Smith) (2023)

Table of Contents:

Introduction
- The Promise of Personalized Medicine: Revolutionizing Healthcare
- Embracing Your Unique Genetic Blueprint for Optimal Health

Chapter 1: Understanding Your Genetic Foundation
- The Basics of Genetics: DNA, Genes, and Chromosomes
- Genetic Variations: Unraveling the Complexity of Your Genome
- The Role of Genetic and Environmental Factors in Health

Chapter 2: Genetic Testing: Insights into Your Unique Genetic Makeup
- Types of Genetic Testing: From DNA Sequencing to Gene Expression Analysis
- Choosing the Right Genetic Test for Your Needs

- Interpreting Genetic Reports: Understanding the Information Provided

Chapter 3: Pharmacogenomics: Personalized Medication Strategies
- How Genes Influence Drug Response: The Science Behind Pharmacogenomics
- Key Genetic Variants Affecting Drug Metabolism and Efficacy
- Implementing Pharmacogenomics in Clinical Practice: Challenges and Opportunities

Chapter 4: Nutrigenomics: Customizing Your Nutrition for Optimal Health
- Genes and Nutrition: Uncovering the Connection
- Personalized Diets Based on Genetic Profiles: Tailoring Nutrition for Individuals
- Nutrigenomics in Disease Prevention and Management

Chapter 5: Personalized Health Coaching: Empowering Your Journey

- The Role of Health Coaches in Personalized Medicine
- Leveraging Personalized Guidance for Lifestyle Modifications and Behavior Change
- Integrating Health Coaching into Traditional Healthcare Systems

Chapter 6: Ethical Considerations in Personalized Medicine
- Privacy and Security: Safeguarding Genetic Data
- Ensuring Equitable Access to Personalized Medicine
- Ethical Implications of Genetic Manipulation and Enhancement

Chapter 7: Case Studies: Real-Life Examples of Personalized Treatment Success
- Inspiring Stories of Individuals Benefiting from Personalized Medicine
- Unique Challenges and Innovative Approaches in Personalized Care
- Lessons Learned from Successful Personalized Treatment Plans

Chapter 8: Integrating Personalized Medicine into Your Life
- Practical Tips for Applying Personalized Medicine Principles in Daily Life
- Navigating Genetic Information and Making Informed Decisions
- Partnering with Healthcare Providers for Effective Collaboration

Chapter 9: The Future of Personalized Medicine
- Advancements in Genomic Research and Technology
- The Potential Impact of Artificial Intelligence and Machine Learning
- Transforming Healthcare: Envisioning a Personalized Medicine Paradigm

Conclusion
- Embracing Personalized Medicine: Empowering Individuals for Health and Wellness
- The Role of Personalized Medicine in Shaping the Future of Healthcare

6

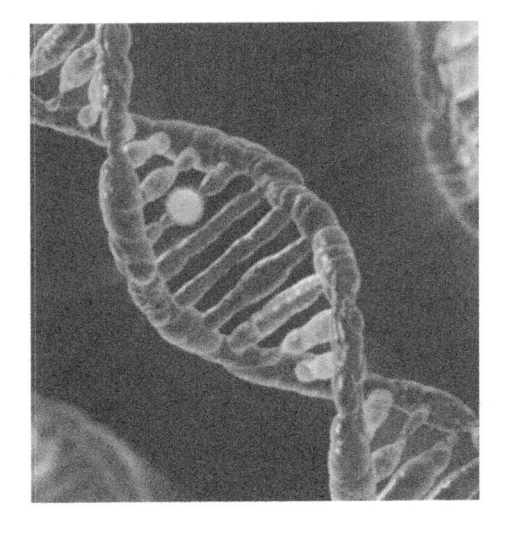

INTRODUCTION

THE PROMISE OF PERSONALIZED MEDICINE: REVOLUTIONIZING HEALTH

In recent years, the field of medicine has witnessed a remarkable transformation driven by groundbreaking advancements in genetic research and technology. This revolution, known as personalized medicine, holds the promise of transforming healthcare as we know it. By harnessing the power of an individual's unique genetic makeup, personalized medicine seeks to tailor medical treatments and interventions to each person's specific needs, optimizing outcomes and improving overall well-being.

Gone are the days of one-size-fits-all approaches to healthcare. With personalized medicine, the focus shifts from a generalized understanding of diseases and treatments to a more individualized and precise approach. It recognizes that each person's genetic composition, lifestyle choices, and environmental factors play a crucial role in determining their health and response to various interventions.

Imagine a healthcare system that not only treats diseases but also prevents them by identifying and addressing underlying genetic predispositions and risk factors. Envision a future where medications are prescribed with precision, taking into account a person's genetic profile to maximize efficacy while minimizing adverse reactions. Picture a world where nutritional plans are tailored to an individual's specific genetic needs, optimizing their health and vitality.

"Genes Unveiled: The Power of Personalized Medicine in Your Hands" takes you on a captivating journey into this realm of personalized medicine. We will delve into the science behind this groundbreaking approach, unraveling the secrets of your unique DNA and exploring how it holds the key to unlocking optimal health and well-being.

Throughout this book, we will explore the different facets of personalized medicine, including pharmacogenomics, nutrigenomics, and personalized health coaching. We will delve into the role of genetic testing, deciphering the information provided in genetic reports, and empowering individuals to make informed decisions about their health.

The field of personalized medicine holds great promise in revolutionizing healthcare by tailoring medical treatments and interventions to individual patients. Traditional approaches to medicine often take a generalized approach, assuming that a particular treatment or medication will have a similar effect on all patients with a given condition. However, it is increasingly recognized that each individual is unique, with variations in genetics, lifestyle, environment, and other factors that influence their response to treatments.

Personalized medicine seeks to unlock this potential by considering the individual characteristics of each patient and tailoring healthcare strategies accordingly. Here are some key aspects that highlight the promise of personalized medicine:

1. Precision Treatment: Personalized medicine allows for precise targeting of treatments based on an individual's unique genetic makeup, biomarkers, and other specific characteristics. By understanding the genetic variants that may influence a person's response to a particular medication, healthcare providers can make more informed decisions about the most effective treatment options. This precision approach can lead to improved treatment outcomes, reduced adverse reactions, and better overall patient experiences.

2. Disease Prevention and Early Detection: Personalized medicine emphasizes proactive measures such as genetic testing, screening, and monitoring to identify individuals at risk of developing certain diseases. By detecting potential health risks at an early stage, interventions can be implemented to prevent the onset or progression of diseases. This shift from reactive to proactive healthcare can lead to significant improvements in patient outcomes and overall population health.

3. Targeted Therapies: Personalized medicine allows for the development of targeted therapies that focus on specific molecular pathways or genetic mutations associated with a particular disease. By identifying the genetic drivers of diseases, researchers and pharmaceutical companies can design therapies that specifically address these underlying causes. This approach has shown promise in various fields, including oncology, where targeted therapies have revolutionized cancer treatment by increasing efficacy and reducing side effects.

4. Pharmacogenomics: Personalized medicine incorporates pharmacogenomics, which explores how an individual's genetic makeup affects their response to medications. By understanding the genetic variations that impact drug metabolism, efficacy, and potential side effects, healthcare providers can optimize medication selection and dosage for each patient. This personalized approach to prescribing medications can lead to improved treatment outcomes, reduced adverse reactions, and enhanced medication safety.

5. Improved Patient Engagement and Experience: Personalized medicine puts the patient at the center of healthcare decision-making, empowering individuals to actively participate in their own health management. By providing patients with personalized information about their genetic predispositions, disease risks, and treatment options, personalized medicine encourages informed decision-making and fosters a collaborative relationship between patients and healthcare providers. This enhanced patient engagement can lead to better treatment adherence, improved health outcomes, and increased patient satisfaction.

6. Data-Driven Healthcare: Personalized medicine relies on the collection and analysis of vast amounts of genetic and health data. As technology advances, the integration of data from various sources, such as electronic health records, genetic sequencing, wearable devices, and lifestyle tracking apps, becomes increasingly feasible. This wealth of data provides insights into disease patterns, treatment responses, and population health trends, enabling healthcare providers and researchers to make evidence-based decisions and develop personalized interventions.

The promise of personalized medicine lies in its potential to transform healthcare from a one-size-fits-all approach to a tailored and individualized approach. By leveraging advancements in genetics, technology, and data analytics, personalized medicine holds the promise of improving treatment outcomes, preventing diseases, enhancing patient experiences, and ultimately revolutionizing the way healthcare is delivered. As research and technology continue to advance, personalized medicine is expected to play an increasingly significant role in shaping the future of healthcare.

By understanding and embracing personalized medicine, you will gain insights into the potential of your genetic code. You will discover how your genes influence medication response, dietary requirements, and lifestyle choices. Armed with this knowledge, you will be equipped to take proactive steps towards improving your health and well-being, with personalized strategies tailored to your unique genetic blueprint.

Join us on this transformative journey as we unveil the power of personalized medicine and its profound implications for revolutionizing healthcare. Together, let us embark on a path that empowers you to take charge of your health, tapping into the remarkable potential hidden within your DNA. Get ready to unlock the secrets of personalized medicine and embark on a life-changing quest towards optimal health and well-being.

EMBRACING YOUR UNIQUE GENETIC BLUEPRINT FOR OPTIMAL HEALTH

Within the intricate fabric of your DNA lies a blueprint that defines who you are. This blueprint holds the key to understanding your predispositions, strengths, and vulnerabilities when it comes to your health. Embracing your unique genetic blueprint is the first step towards unlocking the potential for optimal health and well-being.

In the past, healthcare has often followed a one-size-fits-all approach, where treatments and interventions were based on population averages rather than individual characteristics. However, we now understand that each person's genetic makeup is as distinct as their fingerprints. Your genes influence not only your physical traits but also how your body metabolizes medications, responds to environmental factors, and interacts with your surroundings.

By embracing your unique genetic blueprint, you gain a deeper understanding of your body's inner workings. You can uncover valuable insights into how your genes impact your health, helping you make more informed decisions about your well-being. It empowers you to take a proactive approach, tailoring your healthcare choices to align with your specific genetic profile.

This book, "Genes Unveiled: Personalized Medicine Demystified for Your Health and Well-being," will guide you on this empowering journey. Together, we will navigate the fascinating world of personalized medicine, exploring how your genes can be harnessed to optimize your health.

Through the lens of pharmacogenomics, we will examine how your genetic variations can influence your response to medications. We will explore how this knowledge can guide healthcare professionals in prescribing the most effective and safe treatments, minimizing adverse reactions and maximizing therapeutic benefits.

Nutrigenomics will shed light on the intricate relationship between your genes and nutrition. We will unravel how your genetic makeup influences your dietary needs, allowing you to customize your nutrition for optimal health and vitality. Understanding the genetic nuances of your metabolism and nutrient requirements will empower you to make informed choices that align with your genetic blueprint.

Additionally, we will delve into the realm of personalized health coaching, where guidance tailored to your unique genetic profile can provide invaluable support. Personalized health coaching acknowledges that each individual's journey towards well-being is different. By combining genetic insights with personalized guidance, you can overcome obstacles, adopt positive lifestyle changes, and optimize your health.

By the end of this book, you will have a deeper appreciation for the remarkable potential of personalized medicine. You will recognize the importance of embracing your unique genetic blueprint as a means to unlock the doors to optimal health and well-being. Armed with knowledge and a newfound understanding, you will be equipped to partner with healthcare professionals and take charge of your health journey.

Together, let us embark on this illuminating exploration of personalized medicine. By embracing your unique genetic blueprint, you can pave the way towards a future where healthcare is finely tuned to meet your specific needs. Get ready to unravel the secrets of your DNA, tap into the power of personalized medicine, and embark on a transformative path towards optimal health and well-being.

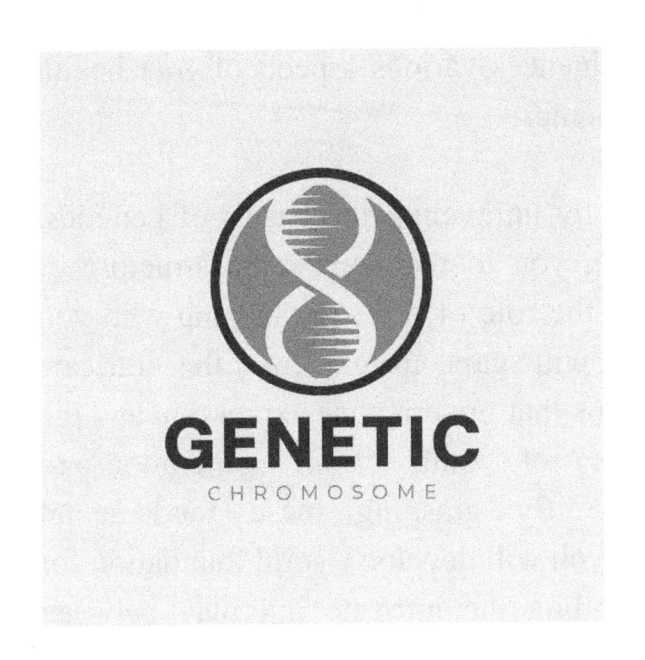

CHAPTER 1: UNDERSTANDING YOUR GENETIC FOUNDATION

In this foundational chapter, we delve into the fascinating world of genetics, providing you with a comprehensive understanding of your genetic foundation. We explore the building blocks of life encoded within your DNA and how it influences various aspects of your health and well-being.

We begin by unraveling the basics of genetics, introducing you to the remarkable structure of DNA and the role of genes in shaping who you are. You will gain insight into the intricate mechanisms that govern gene expression and the inheritance of traits from generation to generation. By grasping these fundamental concepts, you will develop a solid foundation for comprehending the intricate interplay between your genes and your health.

As we journey further, we explore the concept of genetic variations. Every individual carries a unique set of genetic variations that contribute to their distinctive characteristics and health predispositions. We delve into the different types of genetic variations, such as single nucleotide polymorphisms (SNPs), insertions, deletions, and structural variations. Understanding these variations will enable you to appreciate the diversity encoded within the human genome and its implications for personalized medicine.

Building upon this knowledge, we examine the influence of genetic and environmental factors on health. While your genes provide a blueprint, it is essential to recognize that environmental factors can modulate gene expression and impact your health outcomes. We explore the dynamic interplay between genes and the environment, highlighting how lifestyle choices, exposures, and other external factors can interact with your genetic foundation.

Throughout this chapter, we emphasize the importance of genetic testing as a means to gain deeper insights into your genetic foundation. We discuss the different types of genetic tests available, ranging from targeted gene sequencing to whole-genome sequencing. By understanding the potential of genetic testing, you will be empowered to make informed decisions about exploring your genetic makeup and the implications it holds for your health.

Ultimately, this chapter serves as a vital stepping stone in your personalized medicine journey. By understanding your genetic foundation, you will be better equipped to appreciate the profound impact of genetics on your health. Armed with this knowledge, you will embark on a transformative path towards leveraging the power of personalized medicine to optimize your well-being based on your unique genetic blueprint.

Join us in Chapter 1 as we unravel the complexities of your genetic foundation, empowering you to embrace the remarkable potential that lies within your DNA. Get ready to unlock the secrets of your genetic code and embark on a journey towards personalized health and well-being.

SECTION 1: THE BASICS OF GENETICS: DNA, GENES, AND CHROMOSOMES

The scientific study of genes, heredity, and the variation of inherited features in living organisms is known as genetics. Understanding the fundamental building blocks of life and the mechanisms by which traits are passed down from generation to generation is critical to understanding genetics. This section delves into the three fundamental components of genetics: DNA, genes, and chromosomes.

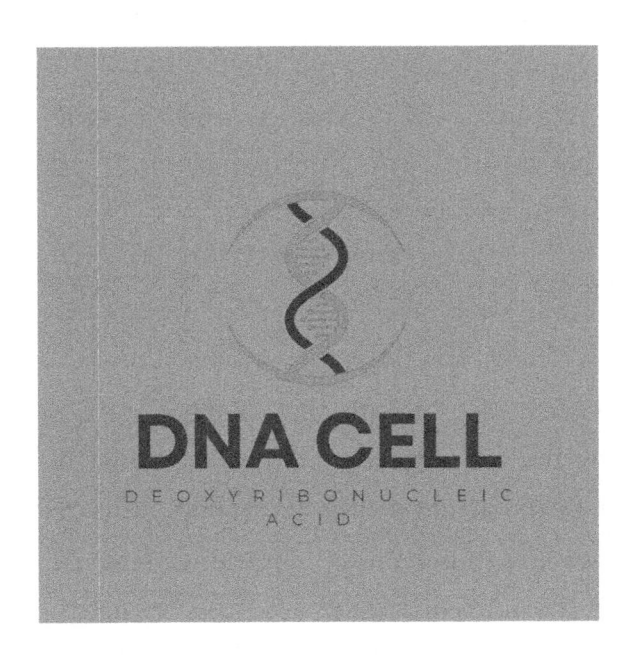

1. Deoxyribonucleic Acid (DNA):

DNA is frequently referred to as the "life blueprint." It's a lengthy molecule made up of nucleotides, which are the building blocks of DNA. Each nucleotide is made up of a deoxyribose sugar molecule, a phosphate group, and one of four nitrogenous bases: adenine (A), thymine (T), cytosine (C), and guanine (G). The arrangement of these bases along the DNA molecule creates the genetic code, which contains instructions for the construction and maintenance of an organism.

2. Genes:

Genes are DNA segments that contain precise instructions for the development of proteins, the functional units in cells that are responsible for diverse biological activities. Genes define our characteristics and attributes by delivering the information required to build proteins that perform certain functions in the body. The size of a gene can range from a few hundred to millions of nucleotides. They are arranged within the DNA molecule in precise regions known as loci. In humans, it is estimated that we possess a range of approximately 20,000 to 25,000 genes within our chromosomes. These genes contain the instructions that determine our traits, characteristics, and susceptibility to certain diseases. By understanding the structure and composition of chromosomes, scientists can unravel the complexities of the human genome and gain a deeper understanding of the genetic basis of life.

3. The Chromosomes:

Chromosomes are intracellular structures that contain DNA tightly coiled around proteins known as histones. When cells divide, these structures can be visualized and studied under a microscope, providing valuable insights into the organization and functioning of our genetic material. Humans have 46 chromosomes, which are grouped into 23 dyads. These pairings include two sex chromosomes (X and Y) that determine a person's biological sex as well as 22 pairs of autosomes. Each chromosome pair contains one chromosome inherited from the mother and one inherited from the father.

Chromosomes play a pivotal part in the transmission of inheritable information from one generation to the coming.. DNA replicates during cell division, and the chromosomes condense, becoming apparent as discrete entities. They guarantee that genetic material is faithfully distributed to daughter cells during cell division, ensuring trait continuity between generations.

4. Inheritance of Genes:

The process through which genetic information is conveyed from parents to offspring is referred to as genetic inheritance. Offspring acquire half of their genetic material from each parent, with the mother inheriting one set of chromosomes and the father inheriting the other. This is accomplished by the creation of gametes (sperm and egg cells) during meiosis, a specialized cell division. During fertilization, the gametes from each parent combine to form a zygote, which grows into an individual with a distinct set of genetic characteristics.

Understanding the fundamentals of genetics, such as DNA, genes, and chromosomes, lays the groundwork for understanding how traits are inherited, genetic variants develop, and genetic information is transformed into the varied characteristics observable in living beings. It serves as the foundation for more sophisticated research in genetics, genomics, and other broad areas of biology.

SECTION 2: GENETIC VARIATIONS: UNRAVELING THE COMPLEXITY OF YOUR GENOME

The human genome is immensely diverse, with tiny variances in genetic sequences contributing to each individual's individuality. These differences, known as genetic variants, are crucial to comprehending the human genome's complexity. This section delves into genetic variants, their origins, and the consequences for human health and traits.

1. SNPs (single nucleotide polymorphisms):

The most prevalent sort of genetic variation in humans is single nucleotide polymorphisms, or SNPs. They involve a single nucleotide base pair mutation inside the DNA sequence. SNPs can be found throughout the genome and can affect many elements of human biology, such as disease susceptibility, treatment reactions, and physical features. The vast majority of SNPs have no discernible impact on an individual's health or traits; however, some SNPs are linked to an increased or decreased risk of specific illnesses or changed physiological responses.

2. CNVs (Copy Number Variations):

Copy number variations are genetic changes caused by the duplication or deletion of a segment of DNA. CNVs can be as small as a few thousand nucleotides or as large as millions of nucleotides, and they can affect entire genes or larger genomic areas. These differences in gene expression and protein production can have a major impact on an individual's phenotype and illness risk. CNVs have been associated with a number of diseases, including developmental abnormalities, cancer, and autoimmune disorders.

3. Deletions and insertions (Indels):

Insertions and deletions, often known as indels, are genetic changes that include the insertion or deletion of a small number of nucleotides in the DNA sequence. Indels can cause frameshift mutations, which change the reading frame and potentially affect gene function. These mutations can have a variety of repercussions, ranging from minor to severe phenotypic impacts, depending on the afflicted gene and the exact indels.

4. Structural Differences:

Structural variants include larger-scale genomic modifications such as chromosomal rearrangements, inversions, translocations, and significant insertions or deletions. These differences can have an impact on the genome's structure and organization, potentially affecting gene regulation, gene expression, and cellular function. Structural differences can contribute to genomic diversity and evolutionary processes, as well as play a role in genetic disorders.

5. Uncommon and Common Variants:

Genetic variations are classified as rare or common based on their prevalence in the population. Rare variations are uncommon and often found in a tiny group of people or families. They are often more likely to be connected with uncommon genetic abnormalities. Common variations, on the other hand, are more common and affect a larger proportion of the population. While each frequent mutation may have a minor impact on features or illness risk, their combined impact can be significant.

Understanding genetic variants is critical for deciphering the human genome's complexity and its impact on health and traits. Researchers can find links between genetic variants and certain disorders or behaviors by investigating and categorizing these variations, leading to advances in personalized medicine, disease prevention, and treatment options. Genetic variants provide the key to unraveling the secrets of our distinct genetic composition, providing vital insights into human biology, evolution, and the intricate mechanisms that govern our
health and well-being.

SECTION 3: THE ROLE OF GENETIC AND ENVIRONMENTAL FACTORS IN HEALTH

A complex interaction of hereditary and environmental variables influences human health. Genetics and the environment both play a role in the development and progression of numerous diseases, as well as in overall health. Understanding these components' involvement is critical for understanding the mechanisms underlying health and disease. This section investigates the roles of genetic and environmental factors in human health.

1. Genetic Variables:

The influence of an individual's genetic composition on their health is referred to as genetic factors. Each person receives a distinct set of genes from their parents, which might influence their susceptibility to specific diseases, reaction to treatments, and overall physiological features. Single nucleotide polymorphisms (SNPs), copy number variations (CNVs), and structural differences can all influence gene expression, protein function, and metabolic pathways, eventually influencing an individual's health outcomes.

Certain genetic variables, such as cancer-related mutations or genetic predispositions to cardiovascular disease, can raise the risk of getting specific diseases. Genetic variables can also influence how a person metabolizes medications, resulting in differences in drug reactions and associated side effects. Furthermore, genetic variables play a role in hereditary illnesses and genetic disorders, which can have a substantial impact on an individual's health and well-being.

2. Environmental Aspects:

Physical, social, and cultural impacts on an individual's health are all examples of environmental factors. These factors can include everything from food, exercise, and cigarette use to pollution, infectious agents, socioeconomic situations, and psychosocial stressors. Environmental influences influence gene expression, epigenetic alterations, and overall physiological function, and they interact with an individual's genetic makeup to influence illness development.

Chronic diseases such as cardiovascular disease, diabetes, and respiratory disorders can be exacerbated by environmental conditions. A bad diet, a lack of physical activity, and exposure to environmental contaminants, for example, can all raise the chance of acquiring obesity and other related disorders. Infectious agents, such as viruses or bacteria, can also cause a variety of infections and contribute to disease development.

3. Interactions between genes and their environments

When genetic factors and environmental exposures interact to influence an individual's health outcomes, this is referred to as a gene-environment interaction. Genetic differences may raise an individual's sensitivity to the adverse impacts of certain environmental conditions in some situations. For example, a genetic predisposition to lung cancer combined with cigarette smoke exposure considerably raises the risk of acquiring the disease.

Gene-environment interactions, on the other hand, can be protective. Environmental factors that are known to enhance illness risk may impart resilience to certain genetic variants. Individuals with certain genetic variants related to cholesterol metabolism, for example, may be less vulnerable to the negative effects of a high-fat diet on cardiovascular health.

4. Epigenetics:

Changes in gene expression patterns that are not induced by changes in the underlying DNA sequence are referred to as epigenetics. Epigenetic changes can be impacted by both genetic and environmental factors, and they play an important role in health and disease. Environmental factors such as nutrition, stress, and toxin exposure can all influence epigenetic marks, resulting in changes in gene expression that can influence disease risk.

Epigenetic changes can be reversed and altered by lifestyle choices, opening up new avenues for illness prevention and control. Researchers are discovering new pathways for customized therapy by better understanding the interplay between genetic and environmental variables in forming the epigenome.

To summarize, both hereditary and environmental variables influence human health and disease. An individual's hereditary susceptibility to diseases and response to treatments are determined by genetic variables. Environmental factors, on the other hand, include all external forces that can have an impact on a person's health. Complex interactions between genetic and environmental factors, such as gene-environment interactions and epigenetic changes, further shape an individual's genetic and environmental profile.

GENETIC TESTING

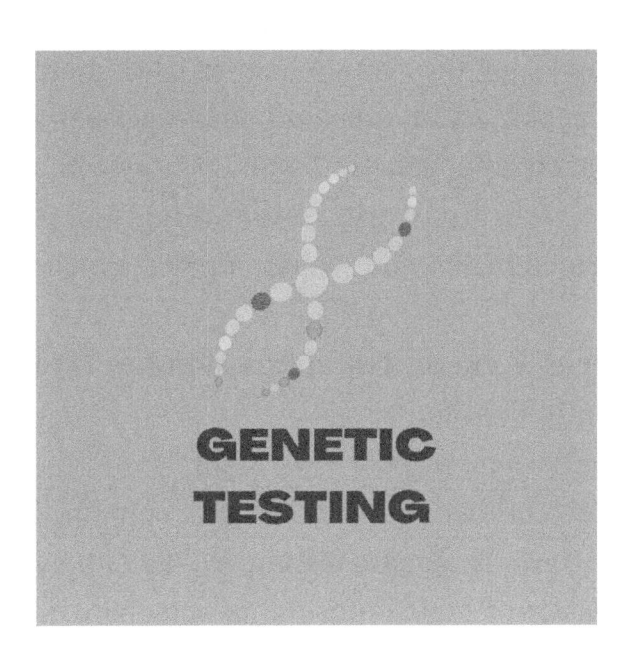

CHAPTER 2: GENETIC TESTING: INSIGHTS INTO YOUR UNIQUE GENETIC MAKEUP

This chapter looks into the field of genetic testing, a powerful instrument that gives us crucial insights into our unique genetic makeup. We look at the various sorts of genetic tests available and how they might help us learn more about our genes, variations, and potential health risks. By accepting genetic testing, we go on a journey of self-discovery, empowering us to make informed health and well-being decisions.

The Power of Genetic Testing: A Window into Your Genetic Profile

The key to unlocking the secrets contained within our genes is genetic testing. We talk about the importance of genetic testing in personalized medicine, emphasizing its ability to deliver relevant knowledge about our genetic background. Through intriguing instances, we demonstrate how genetic testing empowers individuals to take proactive steps toward boosting their health.

Types of Genetic Tests: From Ancestry to Health Insights

We investigate the several genetic tests available, each of which caters to a different component of our genetic composition. We talk about ancestry testing, which gives us a glimpse into our genetic background by deciphering our ancestors' travel patterns and origins. We also look at health-related genetic tests that aim to detect specific genetic changes linked to illness propensity, drug response, and other health-related aspects.

From Sample Collection to Data Analysis: Understanding the Genetic Testing Process

Sample collection, laboratory analysis, and data interpretation are all part of the genetic testing process. We present an overview of the genetic testing process, including how samples are gathered, laboratory techniques used to examine DNA, and data interpretation complexities. Understanding this method allows us to confidently and clearly traverse the world of genetic testing.

Making Informed Decisions: The Benefits and Limitations of Genetic Testing

Genetic testing, like every medical intervention, has advantages and disadvantages. We look at how genetic testing can help you receive unique health insights, uncover potential risk factors, and make proactive healthcare decisions. We also discuss the constraints, such as the likelihood of unreliable results, ethical concerns, and the significance of genetic counseling to aid in result interpretation.

Ethical and Privacy Concerns: Protecting Your Genetic Information
Genetic testing creates significant ethical and privacy concerns. We go into these issues, including the importance of informed consent, genetic data protection, and the potential repercussions of genetic information on personal and familial levels. We can make educated decisions about genetic testing and ensure the proper management and preservation of our

genetic information if we understand these factors.

We obtain important insights into our distinct genetic makeup by accepting genetic testing. The second chapter takes us on a journey through the power of genetic testing, giving us the tools we need to uncover the mysteries contained inside our genes. Join us as we investigate the various types of genetic tests, comprehend the testing procedure, analyze the pros and cons, and manage ethical and privacy issues. Prepare to open the door to self-discovery and embark on a journey toward individualized health and well-being thanks to the invaluable insights supplied by genetic testing.

Gripping the intricate links between genetic and environmental factors allows us to get a comprehensive grasp of the elements that influence our health. In developing individualized healthcare solutions, we realize the importance of both our genetic base and environmental factors. In Section 3, we will investigate the role of genetic and environmental factors in health, decipher the complexities of gene-environment interactions, and pave the way for a breakthrough approach to customized treatment. Prepare to embark on an enthralling adventure in which the dynamic interaction of nature and nurture guides us to optimal health and well-being.

SECTION 1. TYPES OF GENETIC TESTING: FROM DNA SEQUENCING TO GENE EXPRESSION ANALYSIS

DNA sequencing is a game-changing tool that allows scientists to interpret the genetic code and discover the secrets hidden within the blueprint of life. DNA sequencing allows us to grasp the genetic information that shapes the characteristics, traits, and activities of living organisms by establishing the precise order of nucleotides in a DNA molecule. This section delves into the fundamentals of DNA sequencing as well as its applications and impact on numerous disciplines of study and medicine.

1. DNA Sequencing Principles:

At its core, DNA sequencing entails establishing the precise order of the four nucleotides that comprise the DNA molecule—adenine (A), thymine (T), cytosine (C), and guanine (G). The procedure begins with the collection of a DNA sample, which can be obtained from a variety of sources, including cells, tissues, or bodily fluids. The DNA is then split into smaller, more manageable chunks. There are various sequencing techniques, but the most prevalent is known as "next-generation sequencing" (NGS).

NGS makes use of high-throughput sequencing platforms that can sequence millions of DNA fragments at the same time. To establish the order of nucleotides inside each fragment, these platforms use a variety of chemical and enzymatic processes. After that, the sequencing data is examined and rebuilt to produce a whole DNA sequence.

2. DNA Sequencing Applications:

DNA sequencing has transformed multiple fields of study and has far-reaching implications.

Genomics: DNA sequencing has made it possible to sequence complete genomes, including those of humans and many other creatures. This has aided our understanding of genetic variation, evolution, and the molecular foundation of disease.

Medical Diagnostics: DNA sequencing is critical in the diagnosis of genetic illnesses and the identification of disease-causing mutations. It enables the discovery of genetic variants linked to inherited illnesses, allowing for tailored treatment plans and genetic counseling.

Cancer Research: DNA sequencing has transformed cancer research by revealing the genetic mutations and changes that drive tumor formation and progression. It aids in the identification of targeted medicines and the prediction of therapy responses.

Evolutionary Biology: Scientists can untangle evolutionary relationships and examine genetic changes that have occurred over millions of years by comparing DNA sequences from different species.

Forensic Analysis: DNA sequencing is useful in forensic investigations as it aids in the identification of individuals through DNA profiling and the analysis of biological evidence left at crime scenes.

3. Implications for Precision Medicine:

Precision medicine, a novel approach to healthcare, has been made possible by the advent of DNA sequencing. Clinicians can adjust medical treatments and interventions to an individual's unique genetic composition by evaluating their DNA sequence. This tailored strategy improves therapeutic efficacy, reduces side effects, and improves patient outcomes.
DNA sequencing is very useful in pharmacogenomics, which investigates the

genetic basis of pharmacological responses. Clinicians can forecast an individual's response to drugs, modify dosages accordingly, and minimize probable adverse reactions by recognizing unique genetic variants.

4. Future Prospects and Challenges:

Despite the fact that DNA sequencing has changed our understanding of genetics, there are still obstacles to overcome. The interpretation of massive amounts of sequencing data, as well as the integration of genetic information with other clinical and environmental aspects, is a considerable problem. Furthermore, the cost and accessibility of DNA sequencing technology continue to be barriers to widespread genomic medicine implementation.

However, advances in sequencing technologies and bioinformatics are addressing these issues, making DNA sequencing faster, less expensive, and more accessible. This opens the door to more widespread applications in research, diagnostics, and personalized therapy.

To summarize, DNA sequencing is a game-changing technique that has transformed our understanding of genetics and its applications in a variety of sectors. DNA sequencing, by deciphering the blueprint of life, provides vital insights into the genomic basis of traits, diseases, and treatment responses. Its influence on precision medicine and personalized healthcare has enormous promise for improving patient outcomes and furthering medical research. We should expect much larger advances in our understanding of the genetic underpinnings of health and disease as DNA sequencing technology evolves and becomes more accessible.

The capacity to read life's blueprint has created new opportunities for research, diagnostics, and tailored therapy. Scientists are gaining insights into the underlying causes of diseases, identifying novel therapeutic targets, and devising more effective treatment strategies by unraveling the complexity of DNA.

In the field of medical diagnostics, DNA sequencing has given practitioners the ability to make precise and rapid diagnosis of genetic abnormalities. It enables patients and their families to receive early intervention, precise risk assessment, and educated decision-making.

Furthermore, DNA sequencing is important in cancer research because it allows for the discovery of particular mutations that drive tumor growth and guide targeted therapy.

The importance of precision medicine cannot be overemphasized. Clinicians can personalize medicines to an individual's specific needs by assessing their genetic profile, enhancing therapeutic success while reducing unwanted effects. This individualized approach is revolutionizing medicine, resulting in more efficient and effective healthcare delivery.

Despite amazing advances in DNA sequencing, difficulties persist. Large volumes of genetic data require sophisticated bioinformatics tools and knowledge to understand. To enable the proper application of DNA sequencing technology, ethical problems, privacy concerns, and equal access to genomic information and healthcare services must all be addressed.

In the future, DNA sequencing has enormous promise. Continued technological improvements, such as the introduction of third-generation sequencing platforms, will improve sequencing accuracy, speed, and price. Integrating genomic data with other omics data, such as transcriptomics and proteomics, will provide a more complete understanding of human biology's intricacies.

Finally, DNA sequencing has transformed our ability to decode the blueprint of life, with far-reaching consequences for research, diagnostics, and personalized therapy. As we continue to decipher the complexities of the human genome, the impact of DNA sequencing on healthcare will only grow, ushering in a new era of precision medicine in which treatments are tailored to the individual, ultimately leading to improved health outcomes and a higher quality of life for people worldwide.

SECTION 2: CHOOSING THE RIGHT GENETIC TEST FOR YOUR NEEDS

Choosing the Right Genetic Test for Your Needs

Genetic testing is becoming more widely available, and it can provide vital insights into a person's genetic make-up, health concerns, and potential treatment options. However, with so many genetic tests available, it's critical to select the one that best fits your needs and goals. This section investigates the aspects to consider when choosing a genetic test and offers advice on making an informed selection.

1. Determine your goal:
Before selecting a genetic test, you must first determine your objective and what you intend to

obtain from the test. Do you want to learn more about your ancestors and genealogy? Are you concerned about hereditary health issues or your proclivity for particular diseases? Do you wish to look into possible treatments based on your genetic profile? Clarifying your objectives can assist you in narrowing down your selections and selecting a test that meets your specific requirements.

2. Recognize the several types of genetic tests:
There are several sorts of genetic tests available, each with a specific purpose. Here are some examples of common genetic tests to consider:

- Carrier Testing: This sort of test is used to evaluate whether or not a person contains a genetic mutation that could be passed down to their children and cause inherited illnesses.

- Diagnostic Testing: Diagnostic testing is performed when symptoms or indicators of a certain genetic disease are present. These tests aid in the confirmation of a diagnosis and the selection of relevant treatment choices.

- Predictive Testing: Predictive testing determines a person's likelihood of having specific diseases or problems later in life. They can offer insights into genetic predispositions and help inform lifestyle changes or preventative healthcare decisions.

- Pharmacogenomic Testing: Pharmacogenomic testing examines a person's genetic variations to anticipate how they will react to various medications. This information can help healthcare providers adapt pharmacological therapies to enhance effectiveness while minimizing side effects.

- Nutrigenomic Testing: Nutrigenomic testing looks at how a person's genes interact with their food and lifestyle. These tests provide information about tailored nutrition recommendations based on genetic differences.

3. Seek the advice of a healthcare practitioner or genetic counselor.

Genetic testing can have serious consequences for your health and well-being. It is recommended that you consult with a healthcare practitioner or a qualified genetic counselor before making a decision. They can help you understand the advantages, limitations, and potential risks of genetic testing. They will also help you choose the best test for you depending on your medical history, family history, and specific concerns.

4. Assess the test's correctness and dependability:

When contemplating a genetic test, it is critical to evaluate the test's accuracy and reliability, as well as the laboratory doing the study. Look for accredited laboratories that have a great track record in genetic testing and adhere to strict quality standards. Consider the scientific evidence supporting the test's claims as well as the level of validation for the genetic markers examined.

5. Consider the privacy and ethical concerns of the test:

Personal and sensitive information is shared during genetic testing. Before proceeding, thoroughly read the privacy policy of the test provider and understand how they manage and secure your genetic data. Check that the company uses suitable data security measures and that you have given your informed consent to the use and storage of your genetic information.

6. Cost and insurance protection:

The cost of genetic testing can vary greatly based on the type of test and the provider. Consider the financial consequences and whether your insurance covers the test before proceeding. Some tests may be reimbursed by health insurance, especially if a medical indication or a family history of a particular ailment exists. Consult your healthcare practitioner or genetic counselor about insurance coverage and potential out-of-pocket payments.

Finally, selecting the proper genetic test requires careful consideration of your goals, knowledge of the various types of tests available, consultation with healthcare professionals or genetic counselors, and a review of the test's accuracy, privacy measures, and cost. By considering these variables, you can make an informed selection that meets your needs and allows you to get useful insights from genetic testing.

When it comes to understanding your health and well-being, keep in mind that genetic testing is only one piece of the puzzle. It is critical to interpret the results in light of your general medical history, family history, and lifestyle factors. Genetic counseling can assist you in analyzing and comprehending the consequences of your test results.

Genetic testing has the ability to provide tailored information about your genetic make-up, health concerns, and treatment alternatives. It can help you make informed healthcare decisions, implement proactive disease prevention strategies, and even tailor therapies to your specific genetic profile. However, it is critical to approach genetic testing with a clear grasp of your objectives, considerations, and the test's limits.

By following these suggestions and receiving advice from healthcare professionals, you can confidently navigate the process of selecting the proper genetic test. Remember that while understanding your genetic composition can be beneficial, it should always be supplemented by comprehensive medical care and lifestyle choices that contribute to your overall health and well-being.

SECTION 3. INTERPRETING GENETIC REPORTS: UNRAVELING THE INSIGHTS

In this section, we will look at the process of evaluating genetic reports, which can provide vital information on an individual's genetic makeup. Genetic reports include a plethora of information on specific genetic variants and their possible health and disease implications. Individuals and healthcare professionals can uncover the insights contained within genetic data and make informed decisions about personalized health management by learning how to evaluate these reports.

1. SNPs, Insertions, and Deletions in Genetic Variants

Genetic reports frequently include descriptions of several types of genetic variants, such as single nucleotide polymorphisms (SNPs), insertions, and deletions. We investigate the importance of these variations and how they affect gene function and illness risk. Understanding the many sorts of genetic variants addressed in the study is critical for understanding the research's possible consequences for health and personalized medicine.

2. Disease-Associated Variants: Risk and Predisposition Assessment

Specific genetic variations associated with certain diseases or conditions may be highlighted in genetic studies. We examine how these disease-associated variations are discovered and what they represent in terms of determining an individual's risk and proclivity for specific health disorders. Understanding the significance of these polymorphisms can help people take proactive actions toward prevention, early detection, or focused therapies.

3. Pharmacogenomic Variants: Medications Tailored to the Genetic Profile

Pharmacogenomic variants are genetic changes that can affect how a person reacts to various drugs. These polymorphisms may be included in genetic reports, providing insight into how an individual's genetic profile may affect their response to various medicines. We investigate how pharmacogenomic variations can be interpreted and how this knowledge might be used to guide tailored medicine selection and dosage modifications and reduce the risk of adverse drug reactions.

4. Predicting Disease Susceptibility Using Polygenic Risk Scores

Polygenic risk scores (PRS) estimate a person's overall genetic susceptibility to particular diseases. PRS estimations based on the combined effects of various genetic variations may be included in genetic reports. We examine the interpretation and relevance of PRS, as well as its limitations and possible applications in illness risk prediction. Understanding PRS can empower people to adopt lifestyle changes, get appropriate screenings, and participate in early interventions to reduce potential health risks.

5. Genetic Counseling and Consultation: Decision-Making Collaboration

Genetic reports frequently necessitate further discussion and contact with genetically trained healthcare specialists. The need for genetic counseling and consultation in interpreting genetic data and making informed decisions is emphasized. Individuals can share their genetic information, ask questions, address concerns, and explore potential consequences for themselves and their families during genetic counseling sessions. We talk about the collaborative nature of genetic counseling and how it improves the interpretation and comprehension of genetic results.

Understanding genetic variations, illness correlations, pharmacogenomic insights, and the collaborative aspect of genetic counseling are all necessary for interpreting genetic results. Individuals can gain useful knowledge about their genetic makeup, discover potential health hazards, and make informed decisions about individualized health management by deciphering the insights concealed within these reports. This section provides readers with the knowledge and resources they need to effectively navigate genetic reports and apply the insights they provide to their personal well-being.

CHAPTER 3: PHARMACOGENOMICS: PERSONALIZED MEDICATION STRATEGIES

In this chapter, we explore the intriguing field of pharmacogenomics, which investigates how an individual's genetic composition influences their response to medications. Pharmacogenomics offers valuable insights for optimizing medication strategies, reducing adverse drug reactions, and tailoring treatments to maximize their effectiveness. By understanding the principles of pharmacogenomics and its implications for personalized medicine, you can make informed choices about your medication options. We provide an overview of pharmacogenomics, emphasizing its significance in transforming healthcare. We discuss how genetic variations can impact drug metabolism, efficacy, and toxicity, leading to variations in individual responses. By grasping the underlying principles of pharmacogenomics, you will gain an understanding of the potential benefits it holds for personalized medication strategies.

Section 1: The Science behind Pharmacogenomics: How Genes Influence Drug Response

Pharmacogenomics is a field of study that investigates how an individual's genetic makeup affects their reaction to medications. Genes, segments of DNA, contain instructions for creating proteins involved in various biological processes, including drug metabolism, drug targets, and drug transporters. Understanding how genes influence drug response is crucial for developing personalized medication strategies.

1. Drug Metabolism and Enzymes

One way in which genes influence drug response is through drug metabolism. Drug metabolism refers to the processes by which the body breaks down medications into smaller compounds that can be eliminated. Enzymes, proteins encoded by specific genes, play a vital role in drug metabolism. Genetic variations or changes in these genes can impact enzyme activity or expression, resulting in differences in how individuals metabolize and eliminate drugs from their bodies.

For instance, cytochrome P450 (CYP) enzymes are responsible for metabolizing a wide range of medications. Genetic variations in CYP genes can lead to varying enzyme activity levels, affecting drug metabolism. Some individuals may have a rapid metabolism, quickly clearing drugs from their system, while others may have a slow metabolism, resulting in slower drug clearance and potentially higher drug concentrations.

2. Drug Targets and Receptors

Genes also influence drug response by affecting drug targets, specific molecules in the body that drugs interact with to produce therapeutic effects. Genetic variations can alter the structure or function of these drug targets, modifying how drugs bind to them and exert their effects. This can lead to variations in drug efficacy and response among individuals.

For example, a genetic variation in a receptor for a particular drug may impact its binding affinity or sensitivity to the drug. This means that individuals with different genetic variations may require different doses or types of drugs to achieve the desired therapeutic effect. Understanding these genetic variations helps healthcare professionals customize medication regimens for individual patients, maximizing treatment effectiveness.

3. Drug Transporters

Genes also code for drug transporters, proteins involved in the movement of drugs across cell membranes. These transporters play a crucial role in drug distribution and elimination from the body. Genetic variations in the genes responsible for drug transporters can influence transporter activity, resulting in differences in drug absorption, distribution, or elimination.

For instance, genetic variations in drug transporter genes can affect the ability of drugs to enter or exit specific tissues or organs, influencing overall drug efficacy. Understanding these genetic variations helps healthcare professionals optimize drug dosing and selection, ensuring optimal drug distribution and response.

Pharmacogenomics combines the study of genetics, genomics, and pharmacology to uncover how individual genetic variations contribute to differences in drug response. By identifying specific genetic variants that influence drug metabolism, drug targets, and drug transporters, healthcare professionals can tailor medication regimens to individuals, improving treatment outcomes, minimizing adverse effects, and reducing the trial-and-error approach often associated with drug therapy.

Through pharmacogenomic testing and interpretation of genetic information, healthcare providers gain insights into an individual's unique genetic profile, enabling informed decisions regarding medication selection, dosage adjustments, and personalized treatment plans.

SECTION 2: KEY GENETIC VARIANT AFFECTING DRUG METABOLISM AND EFFICACY

1. CYP2D6: The CYP2D6 gene encodes an enzyme responsible for metabolizing a wide range of drugs, including antidepressants, beta-blockers, and opioids. Genetic variations in CYP2D6 can lead to different levels of enzyme activity, resulting in variations in drug metabolism and response. Poor metabolizers may experience reduced drug efficacy, while ultrarapid metabolizers may have a higher risk of adverse drug reactions.

2. CYP2C19: The CYP2C19 gene plays a significant role in the metabolism of drugs such as proton pump inhibitors, antiplatelet agents, and selective serotonin reuptake inhibitors. Genetic variants in CYP2C19 can lead to variations in enzyme activity, resulting in different drug response phenotypes. Poor metabolizers may have reduced drug effectiveness, while ultrarapid metabolizers may experience faster drug metabolism, potentially leading to treatment failure or adverse effects.

3. TPMT: Thiopurine S-methyltransferase (TPMT) is an enzyme involved in the metabolism of thiopurine drugs used to treat conditions like autoimmune disorders and leukemia. Genetic variants in the TPMT gene can result in reduced enzyme activity, leading to increased levels of active drug metabolites. This can increase the risk of drug toxicity and adverse reactions, highlighting the importance of TPMT testing before initiating thiopurine therapy.

4. VKORC1: The VKORC1 gene encodes an enzyme involved in the metabolism of vitamin K, which affects the efficacy of anticoagulant drugs like warfarin. Variations in VKORC1 can impact the required dosage of warfarin to achieve the desired anticoagulation effect. Individuals with certain genetic variants may require lower or higher doses of warfarin to achieve therapeutic levels and reduce the risk of bleeding or clotting events.

5. SLCO1B1: The SLCO1B1 gene encodes a transporter protein responsible for the uptake of statin drugs used to lower cholesterol levels. Genetic variations in SLCO1B1 can affect the efficiency of statin uptake into liver cells, influencing drug efficacy and the risk of side effects. Some variants may result in decreased statin uptake, requiring dose adjustments or the use of alternative statins for optimal cholesterol management.

6. HLA-B: Human leukocyte antigen-B (HLA-B) is a gene involved in immune system regulation. Specific variants of HLA-B are associated with an increased risk of severe adverse reactions to certain medications. For example, the HLA-B*57:01 variant is linked to hypersensitivity reactions to abacavir, a medication used in the treatment of HIV. Genetic testing for HLA-B variants can help identify individuals at risk of such adverse reactions, allowing for personalized medication selection and avoidance of potentially harmful drugs.

Understanding these key genetic variants and their impact on drug metabolism and efficacy is crucial in personalized medicine. Genetic testing can identify individuals with specific variants, enabling healthcare providers to make informed decisions about drug selection, dosing adjustments, and potential risks of adverse reactions. Incorporating this genetic information into clinical practice allows for tailored treatment approaches that optimize therapeutic outcomes, minimize the risk of adverse events, and improve patient safety and satisfaction

SECTION 3. IMPLEMENTATION OF PHARMACOGENOMICS IN CLINICAL PRACTICE: CHALLENGES AND OPPORTUNITIES

The implementation of pharmacogenomics, is the study of how genes influence response to medications, in clinical practice holds great promise for personalized patient care. However, it also presents challenges that need to be addressed for successful integration. This chapter explores the practical aspects of incorporating pharmacogenomics into healthcare settings, highlighting the challenges faced and the opportunities it brings.

Challenges:

1. Education and Training: One of the primary challenges is ensuring healthcare professionals have the necessary education and training to interpret genetic data and effectively apply it in treatment decisions. Understanding the complexities of pharmacogenomics and its clinical implications requires specialized knowledge and ongoing professional development.

2. Integration into Clinical Workflow: Integrating pharmacogenomic testing into existing clinical workflows can be a complex task. It requires rethinking protocols, establishing clear guidelines for test utilization, and incorporating genetic information seamlessly into electronic health records (EHRs) to ensure accessibility and usefulness at the point of care.

3. Data Management and Analysis: Handling and analyzing large-scale genetic data can be challenging. Healthcare systems need robust infrastructure to store, manage, and analyze genetic information securely and efficiently. Additionally, ensuring accurate interpretation of genetic test results and translating them into actionable treatment decisions require sophisticated bioinformatics tools and expertise.

4. Cost and Reimbursement: The costs associated with genetic testing and implementing pharmacogenomics may pose financial challenges for healthcare systems and individuals. Widespread adoption will depend on developing clear reimbursement strategies and demonstrating the value of pharmacogenomic-guided

interventions in improving patient outcomes and reducing healthcare costs.

Opportunities:

1. Personalized Treatment Selection: Pharmacogenomic information enables healthcare providers to tailor drug selection based on an individual's genetic profile. By identifying genetic variants that influence drug response, treatment plans can be customized to maximize therapeutic efficacy and minimize the risk of adverse drug reactions.

2. Improved Drug Safety: Genetic testing can identify patients who are at a higher risk of experiencing adverse drug reactions due to their genetic makeup. Implementing pharmacogenomics allows for proactive measures to prevent harmful drug reactions and enhance patient safety.

3. Enhanced Treatment Outcomes: By considering genetic factors that impact drug response, treatment plans can be optimized to improve patient outcomes. Pharmacogenomics can help identify individuals who are more likely to respond favorably to certain medications, increasing treatment efficacy and reducing treatment failures.

4. Targeted Therapy Development: Pharmacogenomics provides valuable insights into the genetic basis of diseases and drug response, facilitating the

development of targeted therapies. By understanding the specific genetic alterations driving a disease, researchers can design drugs that selectively target those abnormalities, leading to more precise and effective treatment strategies.

Addressing the challenges associated with pharmacogenomics implementation and embracing the opportunities it presents can have a transformative impact on clinical practice. By investing in education and training, streamlining workflow integration, developing robust data management systems, and establishing reimbursement strategies, healthcare systems can unlock the full potential of pharmacogenomics.

The integration of pharmacogenomics into clinical practice has the potential to revolutionize patient care by enabling personalized treatment selection, improving drug safety, enhancing treatment outcomes, and driving targeted therapy development. By proactively addressing the challenges and capitalizing on the opportunities, healthcare systems can pave the way for a future where pharmacogenomics plays a central role in clinical decision-making, benefiting patients worldwide.

CHAPTER 4: NUTRIGENOMICS: CUSTOMIZING YOUR NUTRITION FOR OPTIMAL HEALTH

In this chapter, we delve into the captivating field of nutrigenomics, which explores the intricate connection between our genes, nutrition, and overall well-being. Nutrigenomics focuses on how individual genetic variations influence our response to specific nutrients and dietary components, guiding the customization of nutrition for optimal health. By understanding the principles of nutrigenomics, individuals can make informed dietary choices that align with their unique genetic makeup, enhancing their overall health outcomes.

The Science of Nutrigenomics: Revealing the Genetic-Nutrient Interaction

We begin by unraveling the scientific aspects of nutrigenomics, examining how our genes interact with the nutrients we consume. Nutrigenomics explores how genetic variations affect nutrient metabolism, interactions between nutrients and genes, and subsequent health outcomes. We explore the role of various genetic factors, such as single nucleotide polymorphisms (SNPs), in influencing nutrient absorption, utilization, and the body's response to different dietary components.

Genetic Variations and Nutrient Requirements: Customized Dietary Recommendations

The influence of genetic variations on nutrient requirements is a crucial aspect of nutrigenomics. We delve into how specific genetic variations impact an individual's need for certain nutrients, vitamins, minerals, and other dietary components. Understanding these genetic influences enables personalized dietary recommendations that address an individual's unique nutritional needs, promoting optimal health and preventing nutrient deficiencies or imbalances.

Nutrigenomics and Disease Risk: Uncovering Strategies for Dietary Prevention

Nutrigenomics has shed light on the connection between our genetic makeup, dietary patterns, and disease risk. We explore how certain genetic variations can increase or decrease an individual's susceptibility to various diseases based on their dietary choices. By uncovering these associations, nutrigenomics provides insights into dietary prevention strategies tailored to an individual's genetic predispositions, reducing the risk of developing chronic conditions and promoting long-term health.

Gene-Nutrient Interactions: Exploring Optimal Dietary Combinations

The interplay between genes and nutrients extends beyond individual genetic variations. We examine the concept of gene-nutrient interactions, where specific dietary components can influence gene expression and function. We discuss how certain nutrients can positively or negatively impact gene activity and the potential implications for health. Understanding these gene-nutrient interactions empowers individuals to make informed choices about their dietary habits and the optimal combinations of nutrients to support their genetic well-being.

Nutrigenomics and Personalized Diets: Practical Applications and Future Directions

In this section, we explore the practical applications of nutrigenomics in personalized diets. We discuss emerging technologies, such as genetic testing and analysis, that facilitate the customization of dietary recommendations based on an individual's genetic profile. We also highlight the importance of interdisciplinary collaboration among healthcare professionals, geneticists, nutritionists, and dieticians to ensure the effective integration of nutrigenomics into personalized nutrition plans. Additionally, we discuss the future directions of nutrigenomics research and its potential to revolutionize nutrition and preventive healthcare.

Nutrigenomics offers a groundbreaking approach to nutrition, allowing individuals to tailor their dietary choices based on their unique genetic makeup. By unraveling the genetic-nutrient interaction, understanding the impact of genetic variations on nutrient requirements, and exploring gene-nutrient interactions, we can harness the power of nutrigenomics to optimize our nutrition for optimal health. Implementing nutrigenomics in personalized diets paves the way for precision nutrition, enabling individuals to make informed dietary decisions that align with their genetic predispositions and promote long-term well-being.

SECTION 1. GENE AND NUTRITION: UNCOVERING THE CONNECTION

The relationship between our genes and nutrition is a complex and captivating one. Our genes play a crucial role in determining how our bodies process and respond to the nutrients we consume. They influence our nutrient metabolism, nutrient requirements, and even our susceptibility to certain dietary-related diseases. Understanding this connection is essential for optimizing our dietary choices for optimal health.

1. Genetic Variations and Nutrient Metabolism:

Genetic variations, including single nucleotide polymorphisms (SNPs), can impact how our bodies metabolize and utilize nutrients. These variations can affect the activity of enzymes involved in nutrient metabolism, influencing the absorption, utilization, and elimination of nutrients. For example, certain genetic variations may affect how our bodies process carbohydrates, fats, or proteins, which can impact our energy balance and nutrient needs. By understanding our genetic variations, we can gain insights into our individual nutrient metabolism and make appropriate dietary adjustments.

2. Nutrient Requirements and Genetic Predispositions:

Our genetic makeup can also influence our specific nutrient requirements. Some genetic variations may increase or decrease our need for certain nutrients, vitamins, or minerals. For instance, specific genetic variations may affect the absorption or utilization of vitamin D, iron, or folate, which can impact our overall nutritional status. Identifying these genetic predispositions allows us to personalize our dietary choices to meet our unique nutrient needs and prevent deficiencies or imbalances.

3. Gene-Nutrient Interactions:

The interaction between genes and nutrients extends beyond individual genetic variations. Nutrients can influence gene expression and function through mechanisms like epigenetic modifications. Certain nutrients act as signaling molecules, directly impacting gene activity and affecting our physiological processes. Conversely, gene variations can determine our body's response to specific nutrients. This intricate interplay between genes and nutrients forms the basis of gene-nutrient interactions. Understanding these interactions helps us identify optimal dietary combinations and maximize the benefits of our food choices.

4. Disease Risk and Dietary Influences:

Genetic variations can also influence our susceptibility to dietary-related diseases. Some genetic variations may increase the risk of conditions like obesity, cardiovascular disease, or diabetes when exposed to specific dietary factors. Conversely, other genetic variations may offer protection, reducing the risk of disease even in the presence of less favorable dietary patterns. By uncovering these connections, we can customize our dietary choices to mitigate disease risks and promote long-term health.

5. Personalized Nutrition and Future Directions:

The field of nutrigenomics aims to use our understanding of genes and nutrition to develop personalized nutrition strategies. Advances in genetic testing and analysis enable individuals to obtain information about their genetic variations related to nutrient metabolism, requirements, and disease susceptibility. This information can be used to create personalized nutrition plans that optimize dietary choices based on an individual's genetic profile. Integrating nutrigenomics into clinical practice holds significant potential for precision nutrition, preventive healthcare, and improved health outcomes.

In conclusion, genes and nutrition are intricately linked, influencing our nutrient metabolism, requirements, and disease susceptibility. Understanding this connection through nutrigenomics research allows us to make informed dietary choices that align with our genetic predispositions, optimizing our nutrition for overall health and well-being. By embracing this knowledge, we can harness the power of genes and nutrition to enhance our health and make personalized dietary decisions that promote long-term wellness.

SECTION 2. PERSONALIZED DIETS BASED ON GENETIC PROFILE: TAILORING NUTRITION FOR INDIVIDUALS

The idea that one-size-fits-all in nutrition is no longer applicable. Each person possesses distinct genetic variations that influence how their bodies process and react to various nutrients. Personalized diets, based on genetic profiles, utilize this understanding to customize nutrition recommendations according to an individual's specific genetic composition. By taking into account these genetic variations, personalized diets aim to optimize nutrient intake, prevent deficiencies or imbalances, and enhance overall health and well-being.

1. Genetic Variations and Nutrient Metabolism:

Genetic variations significantly impact how our bodies metabolize and utilize nutrients. These variations can affect the activity of enzymes involved in nutrient metabolism, thereby influencing the efficiency of nutrient absorption, utilization, and elimination. For instance, specific genetic variations can impact the processing of carbohydrates, fats, or proteins, thereby influencing energy balance and nutrient requirements. By analyzing an individual's genetic profile, personalized diets can identify these variations and provide tailored dietary recommendations to support optimal nutrient metabolism.

2. Nutrient Requirements and Genetic Predispositions:

Our genetic makeup can influence our individual nutrient requirements. Certain genetic variations can increase or decrease an individual's need for specific nutrients, vitamins, or minerals. For example, certain genetic variations can affect the absorption or utilization of vitamin D, iron, or omega-3 fatty acids. Understanding an individual's genetic predispositions enables personalized diets to recommend specific foods or supplements to fulfill their unique nutrient needs and prevent deficiencies or imbalances.

3. Dietary Sensitivities and Allergies:

Genetic profiles can also unveil information about an individual's susceptibility to dietary sensitivities or allergies. Certain genetic variations may predispose individuals to food intolerances or allergies, such as lactose intolerance or gluten sensitivity. By identifying these genetic markers, personalized diets can help individuals avoid triggering foods and suggest suitable dietary substitutions to promote digestive health and overall well-being.

4. Optimal Nutrient Combinations and Response:

Personalized diets take into account how an individual's genetic variations influence their response to specific nutrient combinations. Certain genetic markers may indicate an individual's ability to metabolize or utilize certain nutrients more effectively. This information enables personalized diets to recommend optimal nutrient combinations that enhance nutrient absorption, utilization, and health benefits for that individual.

5. Disease Prevention and Management:

Personalized diets based on genetic profiles contribute to disease prevention and management. Specific genetic variations may heighten an individual's risk for particular conditions, such as cardiovascular disease, diabetes, or obesity. By identifying these genetic markers, personalized diets can provide targeted dietary recommendations to mitigate these risks and foster better health outcomes. For individuals with existing health conditions, personalized diets can be tailored to address specific nutritional needs and support disease management.

6. Long-term Wellness and Sustainability:

Personalized diets prioritize long-term wellness and sustainability. By aligning nutrition recommendations with an individual's genetic profile, personalized diets enhance adherence and success rates. These diets consider an individual's food preferences, cultural background, and lifestyle factors, ensuring that dietary changes are realistic and attainable.

In conclusion, personalized diets based on genetic profiles offer an individualized approach to nutrition, considering unique genetic variations and nutritional requirements. By optimizing nutrient intake, addressing sensitivities, and promoting disease prevention and management, personalized diets strive to improve overall health and well-being. Integrating genetic information into dietary recommendations empowers individuals to make informed choices, optimize their nutrition, and embark on a sustainable path to lifelong wellness.

SECTION 3. NUTRIGENOMICS IN DISEASE PREVENTION AND MANAGEMENT

Nutrigenomics, the study of how our genes interact with nutrients, is an evolving field that holds immense potential in disease prevention and management. By delving into the complex relationship between our genetic makeup, nutrition, and disease development, nutrigenomics offers valuable insights into personalized dietary strategies that can help prevent the onset of diseases and optimize management for individuals already affected by specific conditions. Let's explore in more detail how nutrigenomics contributes to disease prevention and management:

1. Understanding Genetic Susceptibility:

Nutrigenomics seeks to understand how specific genetic variations influence an individual's susceptibility to certain diseases. Through genetic profiling, nutrigenomics can identify genetic markers associated with an increased risk of developing particular conditions. This knowledge enables proactive measures to be taken in disease prevention strategies, with a focus on personalized dietary interventions to mitigate the risk. For example, individuals with a genetic predisposition for certain cancers may be advised to follow specific dietary patterns rich in cancer-fighting nutrients.

2. Tailoring Dietary Recommendations:

One of the key applications of nutrigenomics is the provision of tailored dietary recommendations based on an individual's genetic variations. Nutrigenomics takes into account how genetic differences impact nutrient metabolism, utilization, and response. This allows for customized dietary approaches to address specific nutrient requirements and deficiencies, thereby optimizing disease prevention and management. For instance, individuals with genetic variations associated with impaired folate metabolism may benefit from higher folate intake or supplementation to reduce the risk of certain birth defects.

3. Precision Nutrition for Chronic Conditions:

Nutrigenomics offers personalized nutrition strategies for managing chronic conditions. By understanding an individual's genetic variations, nutrigenomics can identify dietary factors that can modulate gene expression and influence disease progression. For example, individuals with genetic variations affecting their lipid metabolism may be advised to follow a diet that emphasizes heart-healthy fats and limits saturated fats to manage cardiovascular disease risk. Precision nutrition approaches, guided by nutrigenomics, can optimize dietary choices to support disease management and enhance overall well-being.

4. Nutritional Interventions and Treatment Efficacy:

Nutrigenomics research investigates how specific nutrients and dietary components interact with an individual's genetic makeup to influence treatment efficacy. By understanding an individual's genetic variations, researchers can identify responders and non-responders to certain treatments or medications. This information helps healthcare professionals customize nutritional interventions that can enhance treatment outcomes and minimize adverse effects. For example, in the field of oncology, nutrigenomics can assist in identifying patients who are more likely to respond to specific chemotherapy drugs and recommend dietary adjustments to optimize treatment efficacy.

5. Lifestyle Modification Support:

Nutrigenomics goes beyond dietary recommendations and extends to insights into lifestyle modifications that can complement disease prevention and management strategies. It takes into account how factors such as physical activity, stress management, and sleep patterns interact with an individual's genetic profile. By tailoring lifestyle recommendations to align with an individual's genetic predispositions, nutrigenomics helps create comprehensive personalized plans for promoting overall health and reducing disease risk. For example, individuals with genetic variations affecting their response to exercise may be advised on specific exercise types or intensities to maximize the benefits.

6. Advancements in Personalized Medicine:

The integration of nutrigenomics with other medical data, such as biomarkers and clinical profiles, contributes to advancements in personalized medicine. By considering an individual's genetic information in conjunction with other relevant factors, healthcare professionals can develop targeted treatment plans that encompass both pharmacological and nutritional interventions. This personalized approach enhances the effectiveness of therapies and improves patient outcomes. Nutrigenomics empowers individuals and healthcare professionals to embrace precision medicine approaches, leading to improved disease management, better health outcomes, and a more personalized and effective approach to overall well-being.

In conclusion, nutrigenomics is a promising field that revolutionizes our approach to disease prevention and management. By unraveling the intricate interplay between genetics and nutrition, personalized dietary strategies can be developed to optimize health outcomes. The integration of genetic information allows for tailored dietary recommendations that address individual nutrient requirements, mitigate disease risks, and enhance treatment efficacy. Nutrigenomics empowers individuals to take control of their health by making informed choices based on their unique genetic profiles. As research in nutrigenomics continues to advance, we can expect further breakthroughs in personalized medicine, precision nutrition, and the overall improvement of individual well-being. By embracing the potential of nutrigenomics, we pave the way for a future where healthcare is increasingly tailored to each person's genetic blueprint, leading to healthier and more fulfilling lives for all.

CHAPTER 5: PERSONALIZED HEALTH COACHING: EMPOWERING YOUR JOURNEY

Personalized health coaching has emerged as an effective and empowering approach to support individuals in their pursuit of optimal health and well-being. This chapter delves into the concept of personalized health coaching, exploring how it can empower individuals to make positive lifestyle changes, improve their overall health, and enhance their quality of life.

Understanding Personalized Health Coaching:

Personalized health coaching involves a collaborative partnership between a trained health coach and an individual who seeks to make positive changes in their health and lifestyle. It surpasses generalized advice and takes into account the unique needs, preferences, and goals of each individual. Health coaches provide guidance, support, and accountability to help individuals navigate their health journey and implement sustainable changes.

Tailoring Strategies to Individual Needs:

One of the greatest strengths of personalized health coaching lies in its ability to tailor strategies to meet the specific needs of each individual. Health coaches consider various factors, such as an individual's health history, current lifestyle habits, preferences, and barriers to change. By understanding these factors, coaches can develop personalized plans that are realistic, achievable, and aligned with the individual's goals.

Goal Setting and Action Planning:

Personalized health coaching places a strong emphasis on goal setting and action planning as essential components of the journey towards improved health. Health coaches work closely with individuals to identify their desired outcomes, whether it involves weight management, stress reduction, better nutrition, or increased physical activity. They assist individuals in breaking down their goals into actionable steps, providing guidance, support, and accountability along the way.

Behavior Change and Sustainable Habits:

At the heart of personalized health coaching lies the process of behavior change. Health coaches assist individuals in identifying and modifying unhealthy behaviors while fostering the adoption of sustainable habits. By addressing underlying beliefs, motivations, and barriers, coaches empower individuals to make long-lasting changes that positively impact their health and overall well-being.

Support and Accountability:

Personalized health coaching creates a supportive environment where individuals feel encouraged, motivated, and accountable for their actions. Health coaches serve as allies, offering guidance, resources, and ongoing support to help individuals stay on track with their health goals. Regular check-ins, progress assessments, and adjustments to the coaching plan ensure that individuals receive the necessary support throughout their journey.

Integrating Technology and Data:

Advancements in technology have enhanced personalized health coaching by providing access to various tools and resources. Mobile apps, wearable devices, and health tracking platforms enable individuals to monitor their progress, track behaviors, and gain valuable insights into their health metrics. Health coaches can leverage these technologies to gather data, provide feedback, and optimize coaching strategies for even greater personalization and effectiveness.

Overcoming Challenges and Sustaining Change:

Personalized health coaching acknowledges that change is a process that can come with challenges and setbacks. Coaches help individuals navigate obstacles, develop resilience, and maintain motivation. They assist in troubleshooting barriers, adapting strategies, and providing ongoing support to ensure sustained progress and long-term success.

In summary, personalized health coaching offers a tailored and empowering approach to improving health and well-being. By considering individual needs, goals, and behavior change, health coaches provide the necessary guidance and support for individuals to make sustainable lifestyle modifications. Through collaboration, accountability, and the integration of technology, personalized health coaching empowers individuals to take control of their health journey, achieve their desired outcomes, and enhance their overall quality of life.

SECTION 1. THE ROLE OF HEALTH COACHES IN PERSONALIZED MEDICINE

Health coaches have a crucial role to play within the realm of personalized medicine, where treatments and interventions are tailored to the individual based on their distinct characteristics, needs, and goals. They serve as a valuable complement to healthcare professionals by offering personalized guidance, support, and accountability to individuals as they navigate their healthcare journey. This section delves into the specific role of health coaches in personalized medicine and elaborates on how they contribute to enhancing patient outcomes.

1. Acting as a Liaison between Patients and Healthcare Providers:

Health coaches serve as a vital bridge between patients and healthcare providers, facilitating effective communication and collaboration. They work closely with individuals to gain a comprehensive understanding of their health concerns, goals, and preferences, and subsequently relay this information to the healthcare team. By acting as a liaison, health coaches ensure that the individual's voice is heard, their needs are addressed, and their treatment plan is tailored to their unique circumstances.

2. Facilitating Behavioral Change:

A primary responsibility of health coaches in personalized medicine is to facilitate behavioral change. They provide guidance and support to individuals as they navigate lifestyle modifications, adhere to treatment plans, and adopt healthy behaviors. Health coaches employ a variety of strategies, such as goal setting, action planning, and motivational interviewing, to empower individuals and bolster their self-efficacy in making sustainable changes.

3. Education and Empowerment:

Health coaches play a pivotal role in educating individuals about their health conditions, treatment options, and self-care strategies. They empower individuals by equipping them with the knowledge and tools necessary to actively participate in their own healthcare. Health coaches adeptly explain complex medical information in a manner that is easily understandable, enabling individuals to make informed decisions and take ownership of their health.

4. Personalized Care Planning:

In the realm of personalized medicine, treatment plans are carefully tailored to each individual's distinct needs, circumstances, and genetic makeup. Health coaches actively contribute to the development of personalized care plans by collaborating with the healthcare team. They offer invaluable insights into the individual's goals, preferences, and lifestyle factors, ensuring that the treatment plan aligns with their values and supports their overall well-being.

5. Support and Accountability:

Health coaches provide ongoing support and accountability to individuals as they navigate their healthcare journey. They serve as a consistent source of encouragement, offering emotional support, guidance, and reassurance. Health coaches help individuals overcome challenges, celebrate successes, and navigate setbacks, fostering resilience and motivation throughout the process.

6. Lifestyle Modification and Disease Prevention:

Within the realm of personalized medicine, lifestyle modification plays a significant role in both disease prevention and management. Health coaches play an instrumental part in guiding individuals towards adopting healthy behaviors, such as engaging in regular exercise, maintaining a balanced nutrition plan, managing stress, and ensuring adequate sleep. They collaborate with individuals to develop personalized strategies and action plans that promote optimal health and mitigate the risk of chronic diseases.

7. Patient Advocacy:

Health coaches function as advocates for individuals within the healthcare system. They ensure that patients' preferences, values, and needs are respected and duly considered in the decision-making process. Health coaches empower individuals to become active participants in their healthcare, guiding them through complex medical systems, providing access to appropriate resources, and assisting in making informed choices that align with their unique circumstances.

In summary, health coaches play a pivotal role in personalized medicine by providing personalized guidance, support, and education to individuals. They facilitate behavioral change, empower individuals to take charge of their health, and foster collaboration between patients and healthcare providers. By bridging the gap between patients and the healthcare system, health coaches contribute to improved patient outcomes, enhanced well-being, and a more patient-centered approach to healthcare delivery.

SECTION 2. LEVERAGING PERSONALIZED GUIDANCE FOR LIFESTYLE MODIFICATIONS AND BEHAVIOR CHANGE

Utilizing personalized guidance is an effective and influential approach when it comes to making lasting lifestyle modifications and achieving behavior change. By taking into account an individual's specific needs, preferences, and circumstances, personalized guidance provides tailored strategies that have a higher likelihood of leading to successful and sustainable outcomes. This section delves into the effective utilization of personalized guidance for lifestyle modifications and behavior change, providing practical steps to maximize its benefits.

1. Gain a Comprehensive Understanding of Your Unique Needs and Goals:

To effectively leverage personalized guidance, it is crucial to gain a clear and comprehensive understanding of your unique needs and goals. Take the time to reflect on your health concerns, desired outcomes, and areas where you would like to make lifestyle modifications. Identify specific behaviors or habits that you would like to change or improve. By defining your goals in detail, you can align the personalized guidance you receive with your aspirations and create a roadmap for success.

2. Seek Support from Qualified Professionals:

To optimize the utilization of personalized guidance, it is important to seek support from qualified professionals who specialize in the specific areas you wish to target. Depending on your individual needs, this may include health coaches, nutritionists, fitness trainers, or therapists. These professionals possess expert knowledge, skills, and experience to provide personalized guidance, strategies, and support that are tailored to your unique circumstances and goals.

3. Conduct a Comprehensive Assessment of Your Current Lifestyle:

Conduct a thorough assessment of your current lifestyle to identify areas that may be contributing to the behaviors you wish to change. Analyze various aspects of your life, such as your daily routine, habits, stress levels, sleep patterns, nutrition, physical activity, and other relevant factors. This self-assessment will help you and your professional support team identify the factors that influence your behaviors, and subsequently develop personalized strategies to overcome challenges and create positive change.

4. Engage in Collaborative Discussions with Professionals:

Active engagement in collaborative discussions with your professional support team is crucial for leveraging personalized guidance effectively. Share your goals, concerns, and any relevant health information with them. This transparent and open communication enables them to gain a comprehensive understanding of your unique circumstances, preferences, and potential barriers to change. Actively participate in the process by asking questions, expressing your needs and concerns, and providing feedback on the strategies proposed by your support team.

5. Develop Actionable Plans:

With the guidance of your professional support team, develop actionable plans that outline specific steps you can take to modify your behaviors and achieve your desired outcomes. These plans should be realistic, achievable, and aligned with your preferences. Break down your goals into smaller, manageable tasks, and set realistic timelines for implementation. Consider incorporating SMART goals (Specific, Measurable, Achievable, Relevant, Time-bound) to provide clarity and track your progress effectively.

6. Regularly Monitor and Track Your Progress:

Consistently keep a check on and track your advancement towards your objectives. Keep a journal, use technology tools, or leverage wearable devices to record your behaviors, achievements, and challenges. Regular monitoring provides valuable feedback and insights into your progress, allowing you to identify patterns, make necessary adjustments, and celebrate your successes along the way. It also enhances accountability and motivation by providing tangible evidence of your efforts.

7. Adapt and Refine Strategies as Needed:

As you navigate your journey of lifestyle modifications and behavior change, it is essential to be open to adapting and refining your strategies. Recognize that change is a process that may require adjustments along the way. Maintain open and ongoing communication with your professional support team, discussing any challenges, barriers, or changes in circumstances that may impact your progress. Collaboratively explore alternative approaches and make necessary modifications to optimize your success.

8. Embrace a Growth Mindset:

Approach the process of lifestyle modifications and behavior change with a growth mindset.

Embrace the notion that setbacks and challenges are opportunities for learning and growth. Be kind to yourself and practice self compassion throughout the trip. Celebrate your successes, no matter how small, and acknowledge the efforts you put into making positive changes. Stay focused on the long-term benefits of your journey and remain committed to your goals. Embracing a growth mindset allows you to view obstacles as stepping stones to progress, enabling you to persevere and continue striving for positive lifestyle modifications.

In conclusion, effectively leveraging personalized guidance for lifestyle modifications and behavior change involves gaining a comprehensive understanding of your unique needs and goals, seeking support from qualified professionals, conducting a thorough assessment of your current lifestyle, engaging in collaborative discussions with professionals, developing actionable plans, regularly monitoring and tracking your progress, adapting and refining strategies as needed, and embracing a growth mindset throughout the process. By following these steps, you can maximize the benefits of personalized guidance and pave the way for successful and sustainable lifestyle modifications and behavior change. Remember, the journey towards positive change is a continuous process, and with the right support and mindset, you can achieve your desired outcomes and improve your overall well-being.

Section 3. Integrating Health Coaching into Traditional Healthcare Systems
SECTION

The integration of health coaching into traditional healthcare systems has gained recognition as a valuable approach to enhancing patient care and improving health outcomes. This innovative model combines the expertise of healthcare professionals with the personalized guidance and support provided by health coaches, resulting in a holistic and patient-centered approach to healthcare delivery. By integrating health coaching into traditional systems, healthcare providers can unlock a range of benefits and strategies that contribute to better patient outcomes and a more comprehensive approach to care.

1. Enhanced Patient Engagement and Empowerment:

Integrating health coaching into traditional healthcare systems promotes active patient engagement and empowerment. Unlike the traditional model where patients may feel passive in their healthcare journey, health coaching encourages patients to take an active role in managing their health. Health coaches work collaboratively with individuals, building a partnership that empowers patients to make informed decisions, take ownership of their health, and actively participate in their treatment plans. Through personalized guidance, goal setting, and behavior change strategies, health coaches equip patients with the tools and knowledge to become active agents in their healthcare.

2. Improved Health Outcomes:

The integration of health coaching has shown promising results in improving health outcomes. Health coaches provide ongoing support, accountability, and motivation to individuals as they navigate their healthcare journey. By focusing on behavior change, adherence to treatment plans, and promoting healthy lifestyle modifications, health coaching can contribute to better patient outcomes. For example, health coaches can help individuals manage chronic conditions more effectively, reduce the risk of complications, and support long-term behavior change that leads to improved overall well-being.

3. Personalized Care Plans:

Health coaching integrates the expertise of healthcare professionals with personalized guidance tailored to the individual's unique needs. This collaborative approach allows for the development of personalized care plans that take into account not only medical interventions but also lifestyle factors, social determinants of health, and patient preferences. Health coaches work with healthcare providers to ensure that the care plans are comprehensive, aligned with the patient's goals, and support the holistic well-being of the individual. By tailoring care plans to individual needs, health coaching contributes to a more patient-centered and effective approach to healthcare.

4. Bridging Gaps in Care:

Integrating health coaching addresses gaps in care within traditional healthcare systems. Health coaches serve as a vital link between healthcare professionals, patients, and community resources. They address the non-medical aspects of health, such as behavior change, lifestyle modifications, and social support, which are often overlooked in traditional healthcare settings. Health coaches can identify barriers to optimal health and work collaboratively with patients to overcome them. By addressing these gaps, health coaching contributes to a more comprehensive and patient-centered approach to care that considers the broader determinants of health.

5. Collaborative Approach to Healthcare:

Integrating health coaching fosters a collaborative approach to healthcare delivery. Health coaches work as part of an interdisciplinary team, collaborating with healthcare professionals to optimize patient care. They serve as a valuable communication link, relaying patient goals, preferences, and progress to healthcare providers. This collaborative model facilitates shared decision-making, enhances care coordination, and improves patient satisfaction. By fostering effective collaboration among healthcare professionals, health coaching strengthens the overall quality and efficiency of healthcare delivery.

6. Integration of Technology:

The integration of technology plays a significant role in the successful implementation of health coaching within traditional healthcare systems. Electronic health records, telehealth platforms, and mobile health applications can support the seamless exchange of information between healthcare providers and health coaches. These technological tools enable remote monitoring, data collection, and communication, allowing for efficient and effective collaboration in delivering personalized care. By leveraging technology, healthcare systems can enhance the integration of health coaching services and ensure that patients receive timely and coordinated support.

7. Training and Education:

To integrate health coaching into traditional healthcare systems, healthcare professionals may require additional training and education. This can involve workshops, certifications,
or collaborative training programs that equip healthcare professionals with the necessary skills

and knowledge to incorporate health coaching principles into their practice. By investing in the training of healthcare professionals, healthcare systems can ensure the successful integration of health coaching services and maximize the benefits for patients. Training programs can cover topics such as motivational interviewing, behavior change strategies, health education, and communication skills. Additionally, ongoing professional development opportunities can keep healthcare professionals up-to-date with the latest research and best practices in health coaching, further enhancing their ability to provide effective and personalized care.

In conclusion, the integration of health coaching into traditional healthcare systems offers numerous benefits for both patients and healthcare providers. By fostering active patient engagement, improving health outcomes, and developing personalized care plans, health coaching enhances the overall quality of care. Moreover, by bridging gaps in care, promoting a collaborative approach to healthcare, and leveraging technology, healthcare systems can effectively integrate health coaching services and provide comprehensive, patient-centered care. Through training and education, healthcare professionals can acquire the necessary skills to deliver effective health coaching interventions, leading to improved patient satisfaction and better health outcomes. By embracing the integration of health coaching, traditional healthcare systems can transform the way care is delivered, placing the patient at the center and empowering individuals to take charge of their health and well-being.

CHAPTER 6: ETHICAL CONSIDERATIONS IN PERSONALIZED MEDICINE

Personalized medicine, which aims to tailor healthcare to individual patients based on their unique genetic makeup and other factors, brings forth a distinct set of ethical considerations. This chapter delves into the ethical dimensions of personalized medicine, exploring the key considerations that arise within this field.

Privacy and Data Protection:

One of the primary ethical concerns in personalized medicine revolves around protecting patient privacy and responsibly handling genetic and personal health information. As personalized medicine heavily relies on extensive data collection, storage, and analysis, it is crucial to ensure that patients' genetic data is adequately safeguarded against unauthorized access, misuse, or discrimination. Ethical guidelines and robust data protection measures need to be in place to maintain patient privacy and uphold the principle of confidentiality.

Informed Consent:

Informed consent stands as a fundamental ethical principle in personalized medicine. Given the complexity and potential implications of genetic testing and personalized interventions, it is essential that patients are fully informed about the nature, purpose, risks, and benefits of these approaches. Patients should possess a clear understanding of the potential outcomes and limitations of personalized medicine, enabling them to make autonomous decisions regarding their participation. Adequate counseling and effective communication are necessary to ensure that patients provide informed consent based on accurate and understandable information.

Equity and Access:

Ethical considerations in personalized medicine extend to issues of equity and access. As the field progresses, it is crucial to ensure that access to personalized medicine interventions and genetic testing is equitable, affordable, and available to all individuals, regardless of socioeconomic status, race, or geographic location. Efforts must be made to address disparities and ensure that the benefits of personalized medicine reach diverse populations, thereby reducing potential inequities in healthcare outcomes.

Psychological and Emotional Impact:

Personalized medicine may have significant psychological and emotional implications for patients and their families. Genetic information can profoundly impact an individual's sense of identity, personal relationships, and future health prospects. Ethical considerations encompass providing appropriate pre- and post-test counseling to help individuals navigate the potential psychological impact of genetic testing results and personalized interventions. This includes ensuring access to mental health support and resources to address any emotional challenges that may arise during the process.

Responsibility in Genetic Counseling:

Genetic counseling plays a vital role in personalized medicine. Genetic counselors have the responsibility to assist individuals in understanding the implications of genetic testing results, making informed decisions, and coping with the potential emotional impact. Ethical considerations within genetic counseling involve providing accurate and balanced information, managing potential conflicts of interest, and respecting patient autonomy. Genetic counselors must prioritize the well-being and best interests of their clients while upholding professional standards and guidelines.

Transparency and Communication:

Ethical considerations in personalized medicine emphasize the need for transparency and effective communication among healthcare providers, researchers, and patients. Clear and accurate communication about the benefits, limitations, uncertainties, and potential risks of personalized interventions is essential to promote shared decision-making and ensure that patients have realistic expectations. Transparent communication also involves disclosing any conflicts of interest and maintaining open dialogue with patients to address their concerns, questions, and preferences.

Research Ethics:

Personalized medicine heavily relies on research to advance understanding, develop new interventions, and improve patient care. Ethical considerations within personalized medicine research encompass obtaining informed consent, protecting participant privacy, ensuring research integrity, and addressing potential conflicts of interest. Researchers must adhere to ethical guidelines and regulatory frameworks to safeguard the rights and well-being of research participants and maintain public trust.

In summary, ethical considerations within personalized medicine encompass privacy protection, informed consent, equity and access, psychological impact, responsible genetic counseling, transparency and communication, and research ethics. Addressing these considerations is vital to uphold the values of autonomy, beneficence, non-maleficence, and justice in the context of personalized medicine. By doing so, we ensure the responsible and ethical application of this evolving field for the benefit of patients and society as a whole and promote the trust and confidence of patients and the broader public in personalized medicine. As the field continues to advance, it is essential for stakeholders, including healthcare professionals, researchers, policymakers, and regulatory bodies, to actively engage in ongoing discussions and collaborations to address emerging ethical challenges.

Continuous evaluation and adaptation of ethical frameworks and guidelines are necessary to keep pace with the rapid advancements in personalized medicine. Additionally, interdisciplinary collaborations between experts in genetics, bioethics, law, and social sciences can contribute to a comprehensive understanding of the ethical implications and help shape policies and practices that prioritize patient welfare and societal well-being.

Ultimately, by upholding ethical principles and ensuring responsible practices, personalized medicine can maximize its potential to deliver targeted and effective healthcare interventions while respecting individual autonomy, promoting justice and equity, and safeguarding patient privacy. By navigating the ethical considerations of personalized medicine thoughtfully, we can foster an environment that fosters innovation, improves patient outcomes, and brings us closer to a future of truly patient-centered and ethically grounded healthcare.

SECTION 1. PRIVACY AND SECURITY: SAFEGUARDING GENETIC DATA

In the field of personalized medicine, where healthcare interventions are customized based on individuals' unique genetic makeup, ensuring the privacy and security of genetic information becomes a critical ethical concern. This section delves into the ethical considerations involved in protecting genetic data, safeguarding patient privacy, and maintaining data security.

1. Significance of Genetic Data Privacy:
Given the sensitive and distinctive nature of genetic data, preserving its privacy becomes paramount. Genetic information reveals personal health details, genetic predispositions, and familial relationships. Protecting the privacy of genetic data is essential to maintain patient confidentiality, prevent potential misuse, and ensure individuals' control over their own genetic information.

2. Legal and Regulatory Frameworks:
Numerous legal and regulatory frameworks have been implemented to protect the confidentiality and integrity of genetic information. These frameworks, such as the Health Insurance Portability and Accountability Act (HIPAA) in the United States and the General Data

Protection Regulation (GDPR) in the European Union, establish specific guidelines for the acquisition, retention, transfer, and disclosure of genetic data. Failure to adhere to these regulations may result in penalties and sanctions.

3. Informed Consent and Data Sharing:
Obtaining informed consent from individuals is crucial for preserving the privacy of genetic data. Patients must be fully informed about the purposes of data collection, potential risks involved, and how their data will be used and shared. Clear consent mechanisms should be in place to allow individuals to make informed decisions about the use and disclosure of their genetic information.

4. Anonymization and De-identification:
To protect privacy, it is important to anonymize or de-identify genetic data whenever possible. Anonymization involves removing personal identifiers from the data, making it impossible to link the genetic information back to a specific individual. De-identification involves modifying

or removing certain data elements to minimize the risk of re-identification.

5. Secure Data Storage and Transmission:
Genetic data should be securely stored and transmitted to prevent unauthorized access or breaches. Robust encryption methods, access controls, and firewalls should be implemented to protect genetic databases and systems. Data should be stored in secure environments, such as certified data centers, with appropriate safeguards against physical and digital threats.

6. Data Sharing and Collaboration:
Collaboration and data sharing among researchers and healthcare providers are essential for advancing scientific knowledge and improving patient care in personalized medicine. However, data sharing must be conducted responsibly and ethically, with privacy and security measures in place. Agreements, such as data use agreements and data sharing agreements, should be established to define the

terms and conditions of data sharing and protect the interests of data contributors.

7. Transparent Privacy Policies:

Healthcare organizations, research institutions, and providers of personalized medicine should have transparent privacy policies in place. These policies should clearly outline how genetic data is collected, stored, used, and shared. Patients should have access to this information and understand their rights regarding their genetic data. Transparent policies foster trust and enable individuals to make informed decisions about participating in personalized medicine initiatives.

8. Ongoing Monitoring and Compliance:

Privacy and security measures should be regularly monitored, evaluated, and updated to keep up with technological advancements and emerging threats. Compliance with privacy regulations and best practices should be an ongoing effort, involving audits, risk assessments, and staff training to ensure adherence to established privacy and security protocols.

To conclude, safeguarding genetic data is crucial in personalized medicine to protect patient privacy, maintain confidentiality, and ensure data security. By implementing legal frameworks, obtaining informed consent, anonymizing data, ensuring secure storage and transmission, promoting responsible data sharing, establishing transparent privacy policies, and maintaining ongoing monitoring and compliance, stakeholders can effectively uphold the privacy and security of genetic data in the context of personalized medicine.

SECTION 2. ENSURING EQUITABLE ACCESS TO PERSONALIZED MEDICINE

Ensuring equitable access to personalized medicine is essential to promote fairness, justice, and equal opportunity in healthcare. Here are some strategies and considerations for achieving equitable access:

1. Awareness and Education: Increase public awareness and understanding of personalized medicine to ensure that individuals from diverse backgrounds are informed about its potential benefits and how it can improve health outcomes. Education programs should be accessible and culturally sensitive, addressing potential disparities in health literacy and ensuring that everyone has equal access to information.

2. Affordability and Insurance Coverage: Address the financial barriers to accessing personalized medicine by working towards affordable pricing models for genetic testing and personalized interventions. Advocate for insurance coverage that includes genetic testing and personalized treatments, ensuring that cost does not prohibit individuals from benefiting from these advancements.

3. Research and Development: Encourage research and development initiatives focused on developing affordable and accessible personalized medicine technologies and interventions. Support funding for research that specifically addresses health disparities and aims to improve healthcare outcomes for underserved populations.

4. Diversity in Research and Clinical Trials: Promote diversity in research and clinical trials to ensure that the benefits and effectiveness of personalized medicine are studied across diverse populations. Including individuals from different ethnic, racial, and socioeconomic backgrounds in research can help identify potential disparities and tailor interventions to meet the needs of diverse populations.

5. Eliminating Bias and Discrimination: Ensure that personalized medicine practices and interventions are free from bias and discrimination. Address potential disparities in access, treatment, and outcomes by implementing policies and guidelines that promote equitable healthcare delivery.

Encourage diversity and inclusion in healthcare workforce and leadership to ensure that personalized medicine is developed and implemented with a broad perspective.

6. Telemedicine and Remote Care: Leverage telemedicine and remote care technologies to overcome geographical barriers and improve access to personalized medicine, especially for individuals residing in rural or underserved areas. Implement telehealth infrastructure and policies that enable remote consultations, genetic counseling, and access to personalized interventions.

7. Collaborative Partnerships: Foster collaborations between healthcare providers, researchers, policymakers, and community organizations to address access disparities and develop strategies for equitable implementation of personalized medicine. Engage stakeholders from different sectors to ensure a comprehensive and inclusive approach to healthcare delivery.

8. Health Equity Assessments: Conduct health equity assessments to identify and address barriers to access and disparities in personalized medicine. This includes analyzing data on utilization rates, health outcomes, and patient experiences to identify gaps and develop targeted interventions to improve access and outcomes for marginalized populations.

9. Policy and Advocacy: Advocate for policies that promote equity in personalized medicine, including legislation that ensures affordable access, insurance coverage, and non-discriminatory practices. Collaborate with advocacy organizations, professional societies, and policymakers to influence policy decisions and drive systemic change.

10. Continuous Monitoring and Evaluation: Regularly monitor and evaluate the implementation of personalized medicine programs to assess their impact on equity. Collect data on access, utilization, and outcomes across diverse populations to identify areas of improvement and refine strategies for equitable access.

11. Community Engagement and Outreach: Engage with communities to understand their unique needs, concerns, and barriers to accessing personalized medicine. Collaborate with community leaders, organizations, and grassroots initiatives to tailor interventions, educational materials, and awareness campaigns that resonate with diverse populations. This approach ensures that personalized medicine is culturally competent and relevant to the communities it serves.

12. Health Literacy and Language Accessibility: Address health literacy challenges and language barriers by providing clear and understandable information about personalized medicine in plain language. Translate educational materials, consent forms, and patient resources into multiple languages to reach individuals with limited English proficiency. Offer language interpretation services during consultations and genetic counseling sessions to ensure effective communication.

13. Addressing Structural Barriers: Recognize and address the structural barriers that contribute to healthcare disparities. These barriers may include limited access to healthcare facilities, transportation challenges, and socioeconomic factors such as income inequality and housing instability. Collaborate with policymakers and community organizations to develop strategies that address these underlying issues and create an environment conducive to equitable access to personalized medicine.

14. Tailoring Interventions to Population-specific Needs: Recognize that different populations may have unique genetic variations, health risks, and social determinants of health. Tailor personalized medicine interventions to address the specific needs and characteristics of different populations. This may involve developing targeted genetic testing panels, interventions, and treatment algorithms that consider the diverse genetic backgrounds and environmental factors that influence health outcomes.

15. Training and Education for Healthcare Providers: Provide training and education to healthcare providers on cultural competency, implicit bias, and the importance of equitable access to personalized medicine. Enhance healthcare providers' understanding of the social and cultural factors that influence health outcomes, and equip them with the skills to provide personalized care to diverse populations. This training should also emphasize the ethical responsibilities of healthcare providers in promoting equitable access to personalized medicine.

16. Collaboration with Industry and Technology Developers: Foster collaboration between academic institutions, healthcare organizations, and industry partners to develop innovative technologies and interventions that are both effective and affordable. Encourage industry partners to prioritize affordability and equitable access when designing and pricing personalized medicine products and services. Public-private partnerships can facilitate the development and dissemination of affordable technologies and interventions.

17. Monitoring and Addressing Bias in Algorithms: Personalized medicine often relies on algorithms and artificial intelligence for data analysis and treatment recommendations. It is crucial to continuously monitor and address potential biases in these algorithms to ensure fair and equitable outcomes. Regularly assess the performance of algorithms across diverse populations and adjust them to minimize any disparities or biases that may arise.

18. Addressing Trust and Cultural Barriers: Recognize that trust and cultural beliefs play a significant role in healthcare decision-making. Understand the cultural nuances and historical contexts that shape individuals' perceptions of personalized medicine. Engage in culturally sensitive discussions, involve trusted community leaders and influencers, and address concerns about privacy, data security, and potential misuse of genetic information.

19. Empowering Patient Advocacy: Empower patients and advocacy groups to actively participate in discussions and decisions related to personalized medicine. Provide opportunities for patient input in research, policy development, and healthcare system planning. Amplify the voices of individuals from marginalized communities to ensure their perspectives and needs are heard and incorporated into personalized medicine initiatives.

20. Long-term Monitoring of Health Outcomes: Continuously monitor health outcomes and assess the impact of personalized medicine interventions on different populations. Evaluate the effectiveness, disparities, and unintended consequences of personalized medicine approaches to refine strategies and ensure equitable access and positive health outcomes for all.

By implementing these strategies and considerations, stakeholders can work towards eliminating barriers and disparities in access to personalized medicine. This comprehensive approach acknowledges the complex interplay of social, cultural, economic, and technological factors and aims to create a healthcare landscape where every individual has equal opportunity to benefit from the advancements of personalized medicine. Achieving equitable access to personalized medicine is an ongoing process that requires collaboration, continuous evaluation, and a commitment to addressing disparities at every level of the healthcare system.

By raising awareness, improving affordability, promoting diversity, addressing bias and discrimination, leveraging telemedicine, fostering collaborations, advocating for equitable policies, and monitoring outcomes, stakeholders can work together to create a healthcare system that prioritizes fairness, justice, and equal opportunity.

Ultimately, the goal is to ensure that personalized medicine is not only accessible to a select few but becomes an inclusive and transformative approach that benefits individuals from all walks of life. Through these efforts, personalized medicine can contribute to reducing health disparities, improving healthcare outcomes, and promoting a more equitable and just society.

SECTION 3. ETHICAL IMPLICATIONS OF GENETIC MANIPULATION AND ENHANCEMENT

Genetic manipulation and enhancement encompass a range of ethical considerations and implications that need to be further explored and addressed. While these technologies hold the promise of significant advancements, it is crucial to navigate them with caution and ethical responsibility.

1. Informed Consent: The importance of informed consent cannot be overstated in the context of genetic manipulation and enhancement. Individuals or entities involved should have a comprehensive understanding of the risks, benefits, and potential long-term implications associated with these technologies. Ethical frameworks should ensure that informed consent is obtained, respecting the autonomy and well-being of those subjected to genetic modifications.

2. Unintended Consequences: The potential for unintended consequences cannot be overlooked. Before implementing genetic modifications, thorough evaluation and assessment of potential risks are necessary. This includes considering unintended side effects and long-term impacts on biodiversity and ecological systems. Ethical decision-making should be guided by precautionary principles to minimize harm and ensure responsible use of genetic manipulation and enhancement technologies.

3. Equity and Justice: Fairness and equitable access are paramount in the ethical considerations surrounding genetic manipulation and enhancement. If these technologies are only accessible to a privileged few, it can perpetuate existing disparities and create an uneven playing field. Ethical frameworks must take into account the distribution of benefits, ensuring that access to genetic manipulation and enhancement is available to all who could benefit from it, without further entrenching social inequalities.

4. Human Dignity and Natural Order: Philosophical and moral concerns are integral to the ethical discourse surrounding genetic manipulation and enhancement. Altering the genetic makeup of humans or organisms raises questions about the inherent dignity of living beings and the potential disruption of the natural order of life. Ethical considerations should explore these concerns, delving into the boundaries of human intervention and the potential consequences of altering fundamental aspects of life.

5. Genetic Enhancement vs. Treatment: Differentiating between genetic enhancement and genetic treatment is crucial in ethical deliberations. Genetic enhancement aims to go beyond what is considered "normal" to enhance certain traits, while genetic treatment focuses on preventing or treating diseases or disorders. Ethical considerations may differ between these two approaches, as enhancement raises questions about defining and valuing certain traits, while treatment focuses on improving health outcomes and alleviating suffering.

6. Genetic Discrimination: Genetic manipulation and enhancement give rise to concerns about genetic discrimination. It is imperative to address the protection of genetic privacy and non-discrimination, ensuring that individuals are not stigmatized, marginalized, or denied opportunities based on their genetic information. Ethical frameworks should encompass safeguards against genetic discrimination in areas such as employment, insurance, and access to healthcare.

7. Long-Term Sustainability: The long-term sustainability of genetic manipulation and enhancement practices requires careful examination. Assessing the impact on biodiversity, ecological systems, and the potential for unintended consequences over generations is vital. Ethical frameworks should account for the preservation of ecological balance and the responsible stewardship of genetic resources, considering the broader implications of these technologies on the environment and future generations.

8. Public Engagement and Transparency: The ethical implications of genetic manipulation and enhancement necessitate active public engagement and transparency. Decision-making processes should involve diverse perspectives and facilitate public discourse to address concerns, values, and ethical considerations. Transparent practices are essential to maintain public trust and ensure that decisions regarding genetic manipulation and enhancement are made in the best interest of society as a whole.

9. Inter-generational Impact: Genetic manipulation and enhancement can have far-reaching consequences that extend beyond the immediate individuals or organisms involved. Changes made to the genetic makeup can be inherited by future generations, raising ethical questions about the implications and responsibilities associated with altering the genetic heritage of entire lineages. Careful consideration must be given to the potential impact on future generations and the moral obligations that arise from such interventions.

10. Cultural and Value Diversity: Genetic manipulation and enhancement intersect with cultural and value systems, as different societies and communities may have varying perspectives on what constitutes acceptable interventions. Ethical frameworks should respect cultural diversity and engage in dialogue with diverse stakeholders to ensure that decisions about genetic manipulation and enhancement account for a range of perspectives and values.

11. Governance and Regulation: The ethical implications of genetic manipulation and enhancement call for robust governance and regulation. Clear guidelines and oversight mechanisms are needed to ensure responsible and ethical use of these technologies. Regulatory frameworks should be flexible enough to adapt to scientific advancements while upholding ethical principles, safeguarding against misuse, and providing mechanisms for accountability.

12. Professional Responsibility: Healthcare professionals and scientists involved in genetic manipulation and enhancement have a unique responsibility to uphold ethical principles. They should adhere to professional codes of conduct, prioritize the well-being of patients and subjects, and ensure that interventions are based on sound scientific evidence and ethical considerations. Continuous education and training on ethical practices are essential to maintain professional integrity.

13. Global Perspective: The ethical implications of genetic manipulation and enhancement transcend national boundaries. Collaborative efforts among countries and international organizations are needed to establish global standards, promote responsible practices, and address potential disparities in access and implementation. Consideration should be given to the potential impact on developing countries and marginalized populations to avoid exacerbating existing inequities on a global scale.

14. Public Perception and Acceptance: Public perception and acceptance play a vital role in shaping the ethical landscape of genetic manipulation and enhancement. Widespread public education and engagement are necessary to foster understanding, address concerns, and promote responsible discussions about the benefits, risks, and ethical considerations associated with these technologies. Building public trust and ensuring that decision-making processes are inclusive and transparent can help create a more ethical and socially accepted framework.

15. Ethical Frameworks for Research: Research involving genetic manipulation and enhancement requires robust ethical frameworks. Institutional review boards and ethics committees should evaluate research proposals to ensure adherence to ethical principles, including informed consent, protection of research subjects, and the responsible conduct of research. Ethical considerations should be an integral part of the research process from study design to data collection, analysis, and dissemination of findings.

16. Long-Term Monitoring and Evaluation: The ethical responsibility extends beyond the initial implementation of genetic manipulation and enhancement technologies. Long-term monitoring and evaluation are necessary to assess the societal impact, identify any unintended consequences, and make necessary adjustments to ethical frameworks and practices. Regular assessments should involve multidisciplinary experts, stakeholders, and impacted communities to guide ongoing ethical decision-making.

17. Global Dialogue and Collaboration: The ethical implications of genetic manipulation and enhancement require ongoing global dialogue and collaboration among researchers, policymakers, ethicists, and the public. International conferences, forums, and collaborations can facilitate knowledge sharing, promote cross-cultural understanding, and foster consensus on ethical principles and guidelines. Such efforts can lead to more comprehensive and globally harmonized approaches to the responsible use of these technologies.

In summary, the ethical considerations surrounding genetic manipulation and enhancement demand a comprehensive examination of informed consent, unintended consequences, equity, human dignity, natural order, genetic discrimination, sustainability, public engagement, and transparency. Striking a balance between the potential benefits and ethical responsibilities can guide the responsible development and application of these technologies, upholding societal values, individual autonomy, and the well-being of all living beings.

By expanding the exploration of informed consent, unintended consequences, equity, human dignity, natural order, genetic discrimination, sustainability, public engagement, governance, professional responsibility, global perspective, public perception, ethical frameworks for research, long-term monitoring and evaluation, and global dialogue and collaboration, a more nuanced understanding of the ethical landscape surrounding genetic manipulation and enhancement emerges. Taking into account these ethical

CHAPTER 7: CASE STUDIES: REAL-LIFE EXAMPLES OF PERSONALIZED TREATMENT SUCCESS

Chapter 7 delves into the realm of personalized medicine by presenting real-life case studies that showcase the remarkable success of personalized treatment approaches. These case studies provide concrete examples of how personalized medicine has revolutionized healthcare and improved patient outcomes. By examining these real-life scenarios, readers will gain a deeper understanding of the practical application and effectiveness of personalized treatment strategies. These case studies serve as inspiring examples of how personalized medicine can make a significant impact on individual patients' lives.

Case Study 1: Personalized Cancer Treatment

Patient: Sarah Thompson

Medical Condition: Breast Cancer

Background: Sarah Thompson, a 45-year-old woman, was diagnosed with an aggressive form of breast cancer. Traditional treatment options, such as chemotherapy, had limited success due to the tumor's resistance to standard therapies.

Personalized Treatment Approach: Sarah's medical team decided to employ a personalized treatment approach based on her tumor's genetic profile. They conducted genomic sequencing of her tumor cells to identify specific genetic mutations driving cancer growth. The analysis revealed a mutation in the HER2 gene, indicating that Sarah's tumor could be susceptible to targeted therapy.

The medical team prescribed a targeted therapy called Herceptin (trastuzumab), which specifically targets HER2-positive breast cancer cells. Sarah underwent treatment with Herceptin along with chemotherapy tailored to her individual case. The targeted therapy successfully inhibited the growth of HER2-positive cells and improved Sarah's response to chemotherapy.

Outcome: Sarah's response to the personalized treatment was remarkable. Her tumor size significantly decreased, and follow-up tests showed no signs of residual cancer cells. Sarah's personalized treatment approach not only improved her overall survival rate but also minimized side effects compared to traditional therapies. Her success story demonstrated the potential of personalized medicine in treating aggressive cancers.

Case Study 2: Personalized Immunotherapy for Melanoma

Patient: John Anderson

Medical Condition: Stage IV Melanoma

Background: John Anderson, a 58-year-old man, was diagnosed with stage IV melanoma, a type of skin cancer that had spread to his lymph nodes and distant organs. Conventional treatments, such as surgery and chemotherapy, had limited effectiveness in controlling the progression of his disease.

Personalized Treatment Approach: John's medical team recommended personalized immunotherapy based on his tumor's genetic profile. They performed genetic testing to identify specific mutations in his tumor cells and discovered a high mutational burden, indicating a potential for a strong immune response against cancer cells.

Based on the genomic analysis, John received a combination of immune checkpoint inhibitors, including pembrolizumab and ipilimumab, to unleash his immune system's ability to target and destroy cancer cells. The treatment was tailored to his unique genetic markers, maximizing the likelihood of a positive response.

Outcome: The personalized immunotherapy approach yielded remarkable results for John. Over time, his tumors began to shrink, and subsequent scans showed significant reduction in tumor size. John experienced a durable response to the treatment, with his cancer going into remission. Personalized immunotherapy offered a promising avenue for advanced melanoma treatment, demonstrating the potential of tailored approaches to harness the power of the immune system in fighting cancer.

Case Study 3: Personalized Gene Therapy for Rare Genetic Disorder

Patient: Emma Johnson

Medical Condition: Spinal Muscular Atrophy (SMA)

Background: Emma Johnson, a 2-year-old girl, was diagnosed with spinal muscular atrophy (SMA), a rare genetic disorder that affects muscle control and movement. SMA is caused by a mutation in the SMN1 gene, resulting in a deficiency of the survival motor neuron (SMN) protein essential for motor neuron function.

Personalized Treatment Approach: Emma's medical team recommended a personalized gene therapy approach called Zolgensma. Zolgensma is a one-time infusion that delivers a functional copy of the SMN1 gene to replace the faulty gene. Prior to treatment, Emma underwent genetic testing to confirm the specific gene mutation causing her SMA and to ensure her eligibility for Zolgensma.

Emma received the gene therapy infusion, and the functional SMN1 gene was successfully integrated into her cells, enabling them to produce the necessary SMN protein. The personalized gene therapy aimed to restore SMN protein production in Emma's cells, which was crucial for the proper functioning of motor neurons. By integrating the functional SMN1 gene into her cells, the therapy sought to address the underlying genetic defect responsible for SMA. The objective was to enable Emma's cells to produce the SMN protein at sufficient levels, promoting the survival and functionality of motor neurons and ultimately improving her muscle control and movement abilities.

The successful integration of the functional SMN1 gene into Emma's cells was a critical step in the personalized gene therapy process. This step aimed to provide a long-term solution by correcting the genetic abnormality causing SMA rather than solely managing its symptoms. By restoring SMN protein production, the therapy aimed to compensate for the deficiency and facilitate the development and maintenance of healthy motor neurons in Emma's body.

The targeted delivery of the functional gene and its successful integration into the cells required precise and sophisticated techniques. Specialized vectors, such as viral vectors, were used to deliver the therapeutic gene to Emma's cells. These vectors were engineered to carry the functional SMN1 gene specifically to motor neurons, ensuring its integration in the appropriate cells and minimizing the potential for off-target effects.

The personalized gene therapy aimed to provide Emma with a treatment approach tailored to her specific genetic condition. By addressing the root cause of SMA, the therapy had the potential to bring about significant improvements in Emma's motor function and overall quality of life. The success of this personalized treatment hinged on the accurate targeting and delivery of the therapeutic gene, as well as the successful integration of the functional gene into her cells.

Following the gene therapy infusion, the newly integrated functional SMN1 gene would enable Emma's cells to produce the SMN protein, which was essential for motor neuron function and overall muscle control. This personalized approach held the promise of long-term benefits, potentially halting or slowing down the progression of SMA and allowing Emma to lead a more independent and fulfilling life.

Outcome: Emma's response to the personalized gene therapy was remarkable. Following the treatment, her motor function began to improve gradually. Over time, Emma gained strength and mobility, reaching developmental milestones that were previously out of reach. Regular assessments and follow-up examinations showed sustained improvement in her muscle control and movement.

Emma's case exemplified the transformative potential of personalized gene therapy in treating rare genetic disorders. By addressing the underlying genetic cause, this tailored approach offered Emma a chance at a better quality of life and the opportunity to live without the limitations imposed by SMA.

These real-life case studies demonstrate the power and success of personalized treatment approaches in diverse medical contexts. Whether it's tailoring cancer therapies based on genetic profiles, leveraging the immune system to combat advanced melanoma, or delivering personalized gene therapy for rare genetic disorders, personalized medicine has shown significant potential in improving patient outcomes, enhancing treatment efficacy, and minimizing side effects. As medical research and technology continue to advance, personalized treatment approaches hold great promise for the future of healthcare, offering tailored solutions to individual patients based on their unique genetic makeup and medical needs.

SECTION 1. INSPIRING STORIES OF INDIVIDUALS BENEFITING FROM PERSONALIZED MEDICINE

1. Sarah's Victory Against Cancer:
Sarah received a diagnosis of an aggressive type of breast cancer. Conventional therapies showed limited efficacy, leaving her with dwindling alternatives. However, personalized medicine offered a glimmer of hope. Through genetic testing, doctors identified specific mutations in her tumor cells. This information guided the selection of targeted therapies tailored to her unique genetic profile. The personalized treatment approach resulted in a remarkable response, leading to tumor shrinkage and improved quality of life. Sarah's story demonstrates how personalized medicine can provide new avenues of treatment and offer hope for patients facing challenging circumstances.

2. Mark's Journey to Mental Health Recovery:

Mark had struggled with depression and anxiety for years, trying various medications with limited success. Personalized medicine offered a different approach. Through genetic testing, doctors discovered genetic variants that influenced Mark's response to different antidepressants. Armed with this knowledge, they were able to identify the most effective medication for him, significantly improving his symptoms and overall mental well-being. Mark's story showcases how personalized medicine can transform mental health treatment by tailoring interventions to an individual's genetic makeup.

3. Emily's Triumph Over a Rare Disease:
Emily was born with a rare genetic disorder that caused severe neurological symptoms. Traditional treatments had limited efficacy, leaving her and her family feeling helpless. However, personalized medicine provided a breakthrough. By conducting whole-genome sequencing, researchers identified the specific genetic mutation responsible for Emily's condition. This knowledge led to the

development of a targeted therapy designed to correct the underlying genetic abnormality. With personalized treatment, Emily experienced a remarkable improvement in her symptoms, offering new possibilities for individuals with similar rare diseases.

4. Jack's Journey to Weight Loss Success:
Jack had struggled with weight management for years, trying various diets and exercise regimens without lasting results. Personalized medicine offered a different approach by analyzing his genetic predispositions related to metabolism and nutrient processing. Through genetic testing, healthcare providers identified specific genetic variations that influenced Jack's response to different types of diets and exercise. Armed with this information, a personalized nutrition and exercise plan was developed to align with Jack's genetic profile. With the tailored approach, Jack experienced significant weight loss and was able to maintain his progress long-term. His story highlights how personalized medicine can

optimize lifestyle interventions for sustainable weight management.

These inspiring stories demonstrate the transformative power of personalized medicine in improving outcomes for individuals facing various health challenges. By leveraging genetic information and tailoring treatments to individuals' unique characteristics, personalized medicine offers new hope, improved outcomes, and enhanced quality of life for patients across a range of medical conditions.

SECTION 2. UNIQUE CHALLENGES AND INNOVATIVE APPROACHES IN PERSONALIZED CARE

Personalized healthcare encounters specific difficulties, but it also presents opportunities for inventive solutions to tackle these obstacles. Let's delve into some of the primary challenges and innovative responses in personalized care:

1. Integration and Management of Data:

A challenge in personalized care involves the integration and management of extensive patient data, encompassing genetic information, electronic health records, and lifestyle data. Innovative approaches entail the development of sophisticated data management systems and technologies capable of securely storing, analyzing, and integrating diverse datasets. This enables healthcare providers to access comprehensive patient information, make informed decisions, and deliver personalized care based on a comprehensive understanding of each individual.

2. Ethical and Legal Considerations:
Personalized care raises ethical and legal concerns concerning genetic privacy, informed consent, and responsible use of genetic information. Innovative solutions involve the establishment of robust ethical frameworks and policies to safeguard patient privacy and ensure transparent and responsible utilization of genetic data. Furthermore, ongoing discussions and collaborations among healthcare professionals,

researchers, policymakers, and patient advocates help address these ethical and legal challenges and shape guidelines for personalized care practices.

3. Collaborative Interdisciplinary Efforts:
Personalized care necessitates collaboration among healthcare professionals from various disciplines, including genetics, genomics, medicine, nursing, and psychology. Collaboration can be challenging due to differing professional backgrounds, knowledge domains, and communication styles. Innovative approaches focus on fostering interdisciplinary collaboration by developing specialized teams, joint education and training programs, and effective communication channels. These collaborative efforts promote a comprehensive and holistic approach to personalized care, ensuring the integration of expertise from multiple disciplines.

4. Education and Training:

Implementing personalized care requires healthcare professionals to acquire new knowledge and skills related to genetics, genomics, and personalized medicine. The challenge lies in providing adequate education and training opportunities to ensure healthcare providers are equipped with the necessary competencies. Innovative approaches involve integrating genetics and genomics education into healthcare curricula, establishing specialized training programs, and utilizing technology-enabled learning platforms. These initiatives empower healthcare professionals to stay updated with the latest advancements in personalized care and deliver evidence-based, patient-centered treatments.

5. Cost and Accessibility:
Personalized care can be associated with higher costs due to genetic testing, specialized treatments, and individualized interventions. This poses challenges in terms of cost-effectiveness and accessibility for all patients. Innovative approaches include

developing cost-effective genetic testing technologies, expanding insurance coverage for personalized care services, and implementing health policies that prioritize equitable access to personalized medicine. These initiatives aim to reduce barriers and ensure that personalized care is accessible to individuals from diverse socioeconomic backgrounds.

In summary, personalized care presents unique challenges regarding data integration, ethics, collaboration, education, cost, and accessibility. However, through innovative approaches such as advanced data management systems, ethical frameworks, interdisciplinary collaboration, education and training programs, and initiatives to improve cost-effectiveness and accessibility, these challenges can be addressed. By overcoming these hurdles, personalized care can continue to evolve and enhance, offering patients tailored and effective treatments that improve outcomes and revolutionize the healthcare landscape.

SECTION 3. LESSONS LEARNED FROM SUCCESSFUL PERSONALIZED TREATMENT PLANS

Successful experiences with personalized treatment plans have yielded invaluable insights that can shape the future of personalized medicine. Here are some key lessons learned from these successful treatment plans:

1. Significance of Genetic Information:

Successful personalized treatment plans highlight the crucial role of genetic information in guiding treatment decisions. By examining an individual's genetic composition, including specific gene variants and mutations, healthcare professionals can identify targeted therapies that are more likely to succeed. These plans emphasize the importance of integrating genetic testing and analysis into clinical practice to optimize treatment outcomes.

2. Individualized Approach:

Personalized treatment plans underscore the necessity of tailoring interventions to the unique characteristics of each individual. Recognizing that no two individuals are exactly alike, successful plans take into account factors such as genetic variations, lifestyle, environmental influences, and personal preferences when designing treatment strategies. By adopting an individualized approach, healthcare providers can better address the specific needs of each patient and enhance treatment outcomes.

3. Collaboration and Multidisciplinary Care: Collaboration among healthcare professionals from various disciplines is a common feature of successful personalized treatment plans. This multidisciplinary approach enables a comprehensive assessment and management of patients' healthcare needs. Geneticists, oncologists, pharmacologists, nutritionists, psychologists, and other specialists collaborate to develop a holistic treatment plan that addresses the complexities of the individual's condition. This collaborative effort enhances the integration of diverse expertise, leading to more effective and comprehensive care.

4. Continuous Monitoring and Adjustments:
Successful personalized treatment plans recognize the importance of continuous monitoring and making adjustments throughout the treatment process. Regular monitoring of patient response, including genetic markers, biomarkers, and clinical outcomes, allows healthcare providers to assess the effectiveness of the personalized treatment and make necessary adjustments. This iterative process ensures that the treatment plan remains aligned with the individual's changing needs and maximizes the chances of success.

5. Patient-Centered Care and Shared Decision-Making:
Patient-centered care and shared decision-making are fundamental aspects of successful personalized treatment plans. These plans prioritize active patient involvement in their care and empower them to participate in

treatment decisions. Patients receive comprehensive information about their condition, available treatment options, and the potential benefits and risks associated with each approach. This shared decision-making process enables patients to make informed choices that align with their values, preferences, and goals.

6. Learning from Failures and Successes:
Successful personalized treatment plans acknowledge the importance of learning from both failures and successes. Failures provide opportunities to reassess treatment strategies, identify areas for improvement, and refine personalized approaches. On the other hand, successes serve as evidence of the potential of personalized medicine and inspire further advancements and innovations. By embracing a learning mindset, healthcare providers can continuously enhance personalized treatment plans and improve patient outcomes.

In conclusion, successful personalized treatment plans have yielded valuable lessons that emphasize the significance of genetic information, individualized approaches, collaboration, continuous monitoring, patient-centered care, and learning from both failures and successes. Incorporating these lessons into future personalized medicine initiatives allows healthcare providers to further optimize treatment outcomes, enhance patient experiences, and advance the field of personalized care.

CHAPTER 8: INTEGRATING PERSONALIZED MEDICINE INTO YOUR LIFE

"Integrate Personalized Medicine into Your Life" refers to the incorporation of personalized medicine principles and practices into an individual's healthcare and lifestyle choices. Personalized medicine focuses on tailoring medical care to the unique characteristics of each person, considering factors such as their genetic makeup, lifestyle, environmental influences, and personal preferences. By integrating personalized medicine into one's life, individuals actively participate in their healthcare and make informed decisions based on personalized treatment approaches.

This may involve undergoing genetic testing, receiving customized treatment plans, adopting lifestyle modifications, and engaging in shared decision-making with healthcare professionals. The goal is to optimize health outcomes, improve treatment efficacy, and enhance overall well-being by leveraging the benefits of personalized medicine. Here are some key considerations and practical strategies to help you integrate personalized medicine into your daily life:

SECTION 1. PRACTICAL TIPS FOR APPLYING PERSONALIZED MEDICINE PRINCIPLES IN DAILY LIFE

Incorporating personalized medicine principles into your daily life can revolutionize the way you approach your health and well-being. By embracing this approach, you can take proactive steps to optimize your health outcomes and make informed decisions about your care. Let's expand on the practical tips mentioned earlier to provide a more comprehensive understanding of how to integrate personalized medicine into your routine:

1. Know Your Family Medical History:
Understanding your family medical history is crucial for identifying potential genetic predispositions and health risks. By delving into your family's health background, you can gain insights into patterns of disease and understand your own risk factors. This knowledge empowers you to have informed conversations with your healthcare provider and develop

personalized strategies for prevention, early detection, and intervention.

2. Seek Genetic Testing and Counseling:
Genetic testing is an invaluable tool for unlocking the secrets hidden within your DNA. By undergoing genetic testing, you can uncover information about your genetic makeup, including specific gene variants and mutations that may impact your health. This knowledge can guide your healthcare decisions, allowing for personalized treatment plans and interventions tailored to your unique genetic profile. Genetic counseling further enhances your understanding of the test results, their implications, and how to effectively incorporate them into your overall healthcare plan.

3. Maintain a Healthy Lifestyle:
A healthy lifestyle is a cornerstone of personalized medicine. It involves making conscious choices regarding nutrition, exercise, sleep, stress management, and other factors that influence your well-being. By customizing your

diet based on personalized recommendations, you can optimize your nutrition to support your genetic profile and individual needs. Engaging in regular physical activity that suits your preferences and capabilities promotes overall fitness and vitality. Prioritizing restful sleep and implementing effective stress reduction techniques contribute to your overall well-being and resilience.

4. Stay Informed and Engage with Healthcare Professionals:

Personalized medicine is an evolving field, with new discoveries and advancements being made regularly. Staying informed about the latest research, breakthroughs, and treatment options empowers you to actively participate in your healthcare journey. Follow reputable sources of information, including scientific journals, reputable websites, and trusted healthcare professionals. Engage in discussions with your healthcare provider, attend seminars or workshops, and join support groups to deepen your understanding of personalized medicine and its application to your specific health needs.

5. Harness the Power of Health Technology:
Health technology offers a range of tools and resources that can support your personalized healthcare journey. Wearable devices, smartphone applications, and digital platforms allow you to monitor and track various aspects of your health, such as physical activity, sleep patterns, heart rate, and nutrition. By leveraging these technologies, you can collect valuable data that provides insights into your health progress, enables early detection of potential issues, and facilitates informed decision-making regarding lifestyle modifications and treatment options.

6. Be Your Own Advocate:
As a participant in personalized medicine, it is crucial to actively advocate for yourself and actively participate in your healthcare decisions. Foster a strong and collaborative relationship with your healthcare provider, sharing your goals, concerns, and preferences openly. Ask

questions, seek clarification, and engage in shared decision-making. Your active involvement ensures that your healthcare plan aligns with your values, priorities, and specific needs. Remember, you are an integral part of the personalized medicine process, and your voice matters.

7. Emphasize Personalized Prevention:
Personalized medicine places significant importance on preventive measures tailored to your individual risks and genetic profile. Stay up-to-date with recommended screenings, vaccinations, and health check-ups that are appropriate for your age, gender, and genetic predispositions. Regularly consult with your healthcare provider to discuss personalized prevention strategies that can minimize your risk of developing certain diseases or conditions. By taking proactive measures, you can detect potential issues early on and take necessary steps to maintain your health and well-being.

By embracing personalized medicine principles and incorporating them into your daily life, you have the opportunity to take control of your health and well-being like never before. This proactive approach allows you to shift from a reactive mindset to a preventive one, addressing potential health risks before they manifest into more significant problems.

Personalized medicine empowers you to make informed decisions based on your unique genetic profile, lifestyle factors, and personal preferences. It recognizes that each individual is different and tailors healthcare strategies accordingly. By following the practical tips mentioned earlier, you can unlock the full potential of personalized medicine and experience its transformative effects.

As you integrate personalized medicine into your routine, keep in mind that it is a continuous journey of learning and adaptation. Stay abreast of the latest advancements in the field, as scientific knowledge and technologies are constantly evolving. Regularly communicate and collaborate with your healthcare provider, ensuring that your personalized healthcare plan remains up-to-date and aligned with your evolving needs.

By embracing personalized medicine, you become an active participant in your own health and well-being. With personalized strategies for prevention, early detection, and treatment, you can optimize your health outcomes and improve your overall quality of life. Embrace this holistic approach and let personalized medicine guide you towards a healthier, more vibrant future.

SECTION 2. NAVIGATING GENETIC INFORMATION AND MAKING INFORMED DECISION

Understanding and navigating genetic information is a multifaceted process that requires careful consideration and guidance. By following the steps outlined below, you can navigate genetic information more effectively and make informed decisions that align with your individual needs and goals.

1. Seek Professional Guidance: Consulting with healthcare professionals who specialize in genetics, such as genetic counselors or medical geneticists, is crucial. These experts possess the knowledge and expertise to help you understand complex genetic information, interpret test results, and provide personalized recommendations. They can address your concerns, answer your questions, and guide you towards making informed decisions.

2. Comprehend the Purpose of Genetic Testing: It is essential to clarify the purpose of genetic testing and what specific information it can provide. Different types of genetic tests offer distinct insights, including disease predispositions, drug responses, or ancestry information. Understanding the purpose and limitations of each test will enable you to set realistic expectations and make decisions that align with your goals.

3. Evaluate the Quality of Genetic Information: Ensure that the genetic information you receive is from reliable and reputable sources. Look for accredited laboratories or companies that adhere to stringent quality standards and follow established guidelines for genetic testing. This ensures the accuracy and reliability of the information you receive, providing a solid foundation for decision-making.

4. Assess Risks and Benefits: Consider the risks and benefits associated with genetic testing and the potential implications of the information you may uncover. Discuss these aspects with your healthcare provider to fully grasp the potential benefits of the information against any psychological, emotional, or privacy-related risks. Understanding the risks and benefits will empower you to make informed decisions that align with your values and priorities.

5. Consider the Impact on Family Members: Genetic information can have implications for your family members as well. Some genetic conditions or predispositions may be hereditary, affecting multiple generations. Discuss with your healthcare provider how the information may impact your family members and the importance of sharing relevant information with them. Open communication within the family can facilitate informed decision-making and proactive management of potential risks.

6. Reflect on Personal Values and Goals: Reflect on your personal values, beliefs, and goals in relation to the genetic information you may uncover. Consider how the information aligns with your priorities and whether it would influence your decisions regarding healthcare, lifestyle choices, or family planning. Understanding your values and goals will enable you to make decisions that are in line with your individual needs and preferences.

7. Make Informed Choices: Armed with the necessary information and having considered the various aspects, make informed choices that align with your goals and values. This may involve incorporating the genetic information into your healthcare plan, making lifestyle changes, or discussing preventive measures with your healthcare provider. Remember that you have the right to make decisions that feel right for you, and seeking additional information or second opinions is always an option.

Navigating genetic information and making informed decisions requires diligence, professional guidance, and an understanding of the potential implications. By following these steps and engaging with healthcare professionals, you can navigate the complexities of genetic information and make decisions that empower you to effectively manage your health and well-being.

SECTION 3. PARTNERING WITH HEALTHCARE PROVIDERS FOR EFFECTIVE COLLABORATION

Establishing and maintaining a productive partnership with healthcare providers is essential for the success of personalized medicine. By following these key steps, you can enhance collaboration and ensure that your healthcare journey incorporates personalized approaches tailored to your unique needs:

1. Foster Open and Transparent Communication: Cultivate a relationship with your healthcare provider based on open and transparent communication. Clearly express your goals, concerns, and expectations related to personalized medicine. Share your interest in exploring tailored healthcare options and discuss how personalized medicine can benefit you. This establishes a collaborative environment where both you and your healthcare provider actively participate in decision-making.

2. Seek Providers with Expertise in Personalized Medicine: Look for healthcare providers who specialize in personalized medicine and have expertise in relevant fields such as genetics, genomics, or pharmacogenomics. Their knowledge and experience will ensure that your healthcare plan incorporates personalized approaches based on your genetic makeup, lifestyle, and other factors. Seek out their guidance and expertise to make informed decisions about your health.

3. Take a Proactive Role in Your Health Journey: Take an active role in your healthcare journey by staying informed about the latest advancements and research in personalized medicine. Educate yourself about your own health and genetic information. Ask questions, seek clarification, and engage in discussions with your healthcare provider. Your proactive approach demonstrates your commitment to personalized care and encourages your healthcare provider to involve you in the decision-making process.

4. Share Complete and Accurate Information: Provide your healthcare provider with a comprehensive overview of your medical history, family history, and any relevant genetic information. Share previous test results, medications, allergies, lifestyle habits, and significant health events. This complete and accurate information enables your healthcare provider to make well-informed decisions and customize your treatment plans to suit your specific needs.

5. Embrace Shared Decision-Making: Embrace a shared decision-making approach with your healthcare provider. Collaborate in discussions about treatment options, medication choices, and lifestyle modifications. Consider the available evidence, weigh the risks and benefits, and take your personal preferences into account. Actively participate in the decision-making process to ensure that your values and goals are considered. This partnership fosters trust, enhances satisfaction, and leads to better healthcare outcomes.

6. Regularly Follow Up and Provide Feedback: Schedule regular follow-up appointments with your healthcare provider to assess the progress of your personalized treatment plan. Share feedback on your experience, including any challenges or successes you have encountered. This ongoing dialogue helps your healthcare provider evaluate the effectiveness of the interventions and make necessary adjustments to optimize your care.

7. Stay Engaged and Proactive: Stay engaged in your healthcare journey beyond clinic visits. Continuously monitor your health, track relevant data, and follow personalized recommendations. Utilize digital health tools, mobile apps, or wearable devices that can facilitate communication with your healthcare provider and provide valuable insights into your health. Proactively communicate any changes or concerns you may have, ensuring that your healthcare provider remains informed and involved in your care.

In conclusion, establishing a productive partnership with healthcare providers is essential for the successful implementation of personalized medicine. By fostering open and transparent communication, seeking providers with expertise in personalized medicine, being proactive in your health journey, sharing complete and accurate information, embracing shared decision-making, regularly following up and providing feedback, and staying engaged and proactive, you can strengthen the collaborative relationship with your healthcare provider. This partnership allows for the integration of personalized approaches based on your unique genetic makeup, lifestyle, and preferences, leading to improved healthcare outcomes and a more tailored and effective healthcare experience. Remember, personalized medicine is a joint effort, and by actively engaging with your healthcare provider, you can optimize the benefits of personalized care and enhance your overall well-being.

CHAPTER 9: THE FUTURE OF PERSONALIZED MEDICINE

The future of personalized medicine holds great promise in revolutionizing healthcare. Advancements in technology, genomics, and data analytics are paving the way for more precise and tailored approaches to individualized care. Here are some key aspects that define the future of personalized medicine:

Precision Diagnostics:

The future of personalized medicine will witness the development of more advanced diagnostic tools. Genomic sequencing techniques will become more affordable and accessible, allowing for comprehensive analysis of an individual's genetic makeup. This will enable earlier detection, accurate disease risk assessment, and personalized treatment selection.

Predictive and Preventive Medicine:

Personalized medicine will increasingly focus on predictive and preventive approaches. By identifying genetic markers and risk factors associated with various diseases, healthcare providers can intervene early and implement preventive measures to minimize the onset or progression of certain conditions. This shift from reactive to proactive healthcare can lead to improved health outcomes and reduced healthcare costs.

Pharmacogenomics and Targeted Therapies:

Pharmacogenomics, the study of how genes influence drug response, will play a pivotal role in personalized medicine. Genetic testing will guide the selection of medications and dosages that are most effective and safe for each individual, minimizing adverse reactions and optimizing treatment outcomes. Targeted therapies, designed to address specific genetic mutations or variations, will become more prevalent, offering highly tailored treatment options.

Incorporation of Artificial Intelligence and Large-Scale Data:

The future of personalized medicine relies on the integration of artificial intelligence (AI) and big data analytics. AI algorithms can analyze vast amounts of patient data, including genomic information, medical records, lifestyle factors, and treatment outcomes, to identify patterns and make predictions. This data-driven approach will enhance treatment decision-making, improve patient risk stratification, and facilitate personalized interventions.

Patient Empowerment and Engagement: Personalized medicine empowers patients to take an active role in their healthcare. With increased access to genetic information and personalized health data, individuals will have a better understanding of their health risks and treatment options. Patient education and engagement will be key in helping individuals make informed decisions and actively participate in their care.

Collaborative Care Models:
Personalized medicine requires collaboration among healthcare providers, researchers, and patients. Multidisciplinary teams, including geneticists, clinicians, pharmacists, and other specialists, will work together to integrate genetic information into clinical practice. Collaborative care models will facilitate the sharing of expertise, improve treatment planning, and ensure comprehensive and personalized patient care.

Ethical Considerations and Privacy Protection:

As personalized medicine evolves, ethical considerations and privacy protection will remain crucial. Safeguarding genetic and personal health information will be paramount, ensuring that individuals have control over their data and how it is used. Striking a balance between data sharing for research and preserving patient privacy will be essential to maintain trust and maximize the benefits of personalized medicine.

In conclusion, the future of personalized medicine holds immense potential to transform healthcare by tailoring treatments to individual patients based on their unique genetic makeup, lifestyle, and environmental factors. By leveraging advancements in technology, embracing data-driven approaches, and fostering collaborative care models, personalized medicine will lead to improved health outcomes, enhanced patient experiences, and a shift towards proactive and preventive healthcare.

SECTION 1. ADVANCEMENTS IN GENOMIC RESEARCH AND TECHNOLOGY

Advancements in genomic research and technology have revolutionized our understanding of the human genome and opened up new possibilities in personalized medicine. Here are some key areas where significant progress has been made:

1. Genomic Sequencing: One of the most significant advancements is the development of high-throughput DNA sequencing technologies. These techniques, such as next-generation sequencing (NGS), enable rapid and cost-effective sequencing of large amounts of genetic material. This has greatly facilitated the decoding of individual genomes and the identification of genetic variations associated with diseases.

2. Precision Medicine: Genomic research has enabled the emergence of precision medicine, which aims to deliver tailored healthcare based on an individual's genetic makeup. By analyzing an individual's genome, researchers can identify specific genetic variants that may influence disease susceptibility, drug response, and treatment outcomes. This allows for the customization of treatments, including the selection of targeted therapies and personalized drug dosages.

3. Genomic Data Analysis: The analysis of genomic data has been greatly enhanced by advancements in bioinformatics and computational biology. Researchers can now analyze vast amounts of genetic information to identify patterns, associations, and potential disease markers. This has led to the discovery of new disease genes, genetic risk factors, and novel therapeutic targets.

4. Functional Genomics: Functional genomics explores the functions and interactions of genes within the context of the entire genome. Techniques such as transcriptomics, proteomics, and metabolomics provide insights into how genes are expressed and regulated, leading to a deeper understanding of disease mechanisms and the development of more targeted therapies.

5. Genome Editing Technologies: The advent of genome editing technologies, such as CRISPR-Cas9, has revolutionized genetic engineering and gene therapy. These tools allow scientists to precisely modify specific genes, correcting disease-causing mutations or introducing therapeutic genes. Genome editing holds promise for treating genetic disorders and has the potential to transform the field of medicine.

6. Single-Cell Genomics: Single-cell genomics enables the study of individual cells, providing insights into cellular diversity and heterogeneity within tissues and organs. This technology allows for the identification of rare cell populations, the characterization of cell states, and the understanding of cellular responses to diseases and treatments. Single-cell genomics has the potential to uncover new biomarkers and therapeutic targets.

7. Data Sharing and Collaborative Research: Genomic research has thrived through the sharing of data and collaborative efforts. Large-scale initiatives, such as the Human Genome Project and international genomic consortia, have facilitated the sharing of genomic data, leading to accelerated discoveries and advancements. Data sharing promotes collaboration, enhances statistical power, and enables researchers to make more comprehensive and impactful findings.

These advancements in genomic research and technology have accelerated our understanding of the human genome and its implications for health and disease. They have paved the way for personalized medicine, offering the potential for improved diagnostics, targeted therapies, and better health outcomes. As technology continues to advance, genomics will play an increasingly integral role in shaping the future of medicine.

SECTION 2. THE POTENTIAL IMPLICATIONS OF ARTIFICIAL INTELLIGENCE AND MACHINE LEARNING

The potential impact of artificial intelligence (AI) and machine learning in various fields, including healthcare, is immense. In the context of personalized medicine, AI and machine learning techniques have the potential to revolutionize healthcare delivery, diagnostics, treatment selection, and patient outcomes. Here are some key areas where AI and machine learning can make a significant impact:

1. Disease Diagnosis and Risk Prediction: AI algorithms can analyze vast amounts of patient data, including medical records, genetic information, imaging results, and lifestyle factors, to identify patterns and make accurate diagnoses. Machine learning models can learn from historical data to predict disease risks and identify individuals who are more likely to develop certain conditions. This can enable early intervention and preventive strategies.

2. Precision Treatment and Drug Discovery: AI can assist in identifying optimal treatment options tailored to an individual's unique genetic profile and medical history. Machine learning algorithms can analyze genomic data to predict drug response, identify potential side effects, and optimize medication dosages. AI can also expedite the drug discovery process by analyzing large datasets and identifying potential therapeutic targets.

3. Medical Imaging and Diagnostics: AI algorithms can analyze medical images, such as X-rays, CT scans, and MRIs, to detect abnormalities and assist in diagnostics. Machine learning models can be trained to identify patterns and features indicative of specific diseases, enabling more accurate and timely diagnosis. AI can also help in automating image interpretation, reducing human error, and improving efficiency.

4. Personalized Treatment Recommendations: AI-powered systems can provide personalized treatment recommendations based on individual patient characteristics, medical history, and clinical guidelines. By considering a wide range of variables and evidence-based guidelines, AI can assist healthcare providers in making informed decisions and tailoring treatments to each patient's specific needs. This can lead to improved treatment outcomes and patient satisfaction.

5. Real-Time Monitoring and Predictive Analytics: AI can enable real-time monitoring of patient data, including vital signs, symptoms, and biomarkers. Machine learning algorithms can detect patterns and trends that may indicate disease progression, adverse events, or treatment response. This facilitates early intervention, timely adjustments to treatment plans, and proactive patient management.

6. Clinical Decision Support Systems: AI-powered clinical decision support systems can provide healthcare providers with evidence-based recommendations, treatment guidelines, and relevant medical literature. These systems can analyze patient data, medical history, and the latest research to support clinical decision-making, reducing errors and improving patient care.

7. Patient Engagement and Education: AI-powered applications and virtual assistants can provide personalized health information, patient education materials, and lifestyle recommendations. These tools can empower individuals to actively manage their health, make informed decisions, and adopt healthy behaviors. AI can also facilitate remote patient monitoring, telemedicine, and virtual consultations, enhancing access to healthcare services.

While AI and machine learning hold tremendous potential, their successful integration into healthcare requires addressing challenges such as data privacy, algorithm transparency, and ethical considerations. Collaborations between healthcare professionals, data scientists, and policymakers are essential to ensure the responsible and ethical use of AI in personalized medicine, while maximizing its benefits for patients and healthcare systems.

SECTION 3. TRANSFORMING HEALTHCARE: ENVISIONING A PERSONALIZED MEDICINE PARADIGM

The shift towards personalized medicine represents a transformative change in healthcare delivery and patient care. This approach harnesses the power of tailoring healthcare interventions to individual patients based on their unique characteristics, genetic makeup, lifestyle, and environmental factors. Here are further expanded key aspects of this transformation:

1. Precision in Diagnosis and Treatment: Personalized medicine places a strong emphasis on precise diagnosis and treatment. By considering individual variations in genetics, biomarkers, and other relevant factors, healthcare providers can move away from a one-size-fits-all approach. Instead, they leverage advanced technologies and comprehensive patient data to identify specific disease drivers, predict disease progression, and determine the most effective and targeted treatment options for each patient.

2. Preventive and Proactive Care: A cornerstone of personalized medicine is the emphasis on preventive and proactive care. Through the analysis of an individual's genetic predispositions, lifestyle choices, and environmental factors, healthcare providers can identify potential health risks at an early stage. This enables the implementation of tailored preventive strategies, empowering individuals to make informed decisions about their health and take proactive measures to maintain well-being and prevent the development of diseases.

3. Integration of Data and Technology: Personalized medicine relies on the integration of diverse data sources, such as electronic health records, genetic information, wearable devices, and patient-reported outcomes. This wealth of data is harnessed through advanced data analytics, artificial intelligence, and machine learning algorithms. By deriving meaningful insights, identifying patterns, and making data-driven decisions, personalized medicine leverages the integration of data and technology to turn the vision of precision medicine into a tangible reality.

4. Collaboration in Healthcare: Personalized medicine advocates for a collaborative approach among healthcare professionals, researchers, patients, and other stakeholders. It recognizes the value of interdisciplinary collaboration, knowledge sharing, and ongoing dialogue to improve patient outcomes and drive scientific discoveries. By fostering collaboration, personalized medicine facilitates the translation of genomic research into clinical practice, ensuring that advancements benefit patients and contribute to the continuous growth of this field.

5. Empowerment and Engagement of Patients: At the core of personalized medicine is the empowerment and engagement of patients in their own healthcare journey. Patients are no longer passive recipients of care but are actively involved in decision-making processes. They are provided with comprehensive information about their health, including their genetic information, and encouraged to participate in personalized treatment plans. This active engagement fosters a patient-centered approach, leading to better adherence to treatment regimens, improved health outcomes, and increased patient satisfaction.

6. Ethical and Regulatory Considerations: The adoption of personalized medicine requires a comprehensive assessment of ethical and regulatory frameworks. As healthcare embraces the power of individualized care, it becomes crucial to address ethical concerns and establish robust regulations. One of the primary considerations is safeguarding the privacy and security of genetic and personal health data. Stricter measures must be in place to ensure that sensitive information is protected and used responsibly, with patients having control over how their data is accessed and shared.

In conclusion, personalized medicine represents a paradigm shift in healthcare, focusing on precision, prevention, integration, collaboration, and patient empowerment. By embracing this approach, healthcare can move towards providing more effective, targeted, and patient-centered care. The continued advancement of personalized medicine holds the potential to revolutionize healthcare and improve the lives of individuals by unlocking the full potential of their unique characteristics and genetic makeup.

CONCLUSION

EMBRACING PERSONALIZED MEDICINE: EMPOWERING INDIVIDUALS FOR HEALTH AND WELLNESS

Embracing personalized medicine has the potential to empower individuals and transform the landscape of health and wellness. By tailoring healthcare interventions to the unique characteristics of each individual, personalized medicine puts individuals at the center of their own care, fostering a sense of empowerment, engagement, and ownership over their health.

One of the key advantages of personalized medicine is its ability to provide individuals with personalized information about their genetic makeup, disease risks, and treatment options. This knowledge empowers individuals to make informed decisions regarding their health and well-being. Armed with this information, individuals can actively participate in discussions with healthcare providers, ask relevant questions, and advocate for personalized treatment plans that align with their values, preferences, and goals.

Embracing personalized medicine also enables individuals to take a proactive approach to their health. By understanding their genetic predispositions, individuals can adopt preventive measures, make lifestyle modifications, and engage in behaviors that promote wellness and disease prevention. Personalized medicine empowers individuals to embrace a healthier lifestyle, including tailored nutrition plans, personalized exercise regimens, and stress management strategies that are specifically tailored to their genetic profile and individual needs.

Furthermore, personalized medicine facilitates early detection and more accurate diagnoses, leading to timely interventions and improved health outcomes. By utilizing genetic testing and advanced diagnostic tools, healthcare providers can identify diseases at their earliest stages and develop personalized treatment plans. This early intervention not only improves treatment efficacy but also enhances individuals' chances of achieving positive health outcomes.

Additionally, embracing personalized medicine fosters a collaborative partnership between individuals and their healthcare providers. Shared decision-making becomes the cornerstone of healthcare interactions, as individuals are encouraged to actively participate in their care plans, voice their concerns, and contribute to treatment decisions. This collaborative approach not only strengthens the patient-provider relationship but also improves treatment adherence and patient satisfaction.

To fully realize the potential of personalized medicine, it is essential to address challenges related to data privacy, equity in access to genetic testing, and the integration of personalized medicine into healthcare systems.

Collaboration among healthcare professionals, researchers, policymakers, and individuals themselves is crucial in overcoming these challenges and creating a healthcare ecosystem that embraces the principles of personalized medicine.

In conclusion, embracing personalized medicine empowers individuals to take charge of their health and well-being. By providing personalized information, promoting proactive behaviors, facilitating early detection, and fostering collaborative partnerships, personalized medicine holds the key to a future where individuals are active participants in their healthcare journey. By embracing personalized medicine, we can revolutionize healthcare, improve health outcomes, and empower individuals to lead healthier and more fulfilling lives.

THE ROLE OF PERSONALIZED MEDICINE IN SHAPING THE FUTURE OF HEALTH-CARE

Personalized medicine holds tremendous potential in shaping the future of healthcare by revolutionizing how we prevent, diagnose, and treat diseases. With advancements in genomics, digital health technologies, and data analytics, personalized medicine is paving the way for more precise, targeted, and effective healthcare interventions. By tailoring treatments to individual characteristics, including genetic makeup, lifestyle factors, and environmental influences, personalized medicine has the ability to improve patient outcomes, enhance patient experiences, and optimize healthcare resource utilization.

One of the key benefits of personalized medicine is its ability to shift the focus from a one-size-fits-all approach to a patient-centered model of care. By considering each individual's unique characteristics, personalized medicine enables healthcare providers to develop tailored treatment plans that address the specific needs, preferences, and goals of patients. This patient-centric approach promotes shared decision-making, empowers patients to actively participate in their healthcare, and ultimately improves treatment adherence and satisfaction.

Moreover, personalized medicine has the potential to revolutionize disease prevention. By identifying genetic predispositions and risk factors through genetic testing, individuals can take proactive measures to mitigate their risk and adopt preventive strategies specific to their genetic profile. This shift towards personalized prevention has the potential to significantly reduce the burden of diseases and improve population health outcomes.

Furthermore, personalized medicine has the capacity to optimize the use of healthcare resources. By tailoring treatments to individuals, healthcare providers can minimize the use of ineffective or unnecessary interventions, reduce adverse effects, and allocate resources more efficiently. This targeted approach not only enhances patient outcomes but also contributes to cost savings and sustainability in healthcare systems.

As personalized medicine continues to evolve, it is crucial to address challenges related to data privacy, ethical considerations, access to genetic testing, and equitable distribution of personalized treatments. Collaboration among healthcare professionals, researchers, policymakers, and patients will be essential in overcoming these challenges and ensuring that the benefits of personalized medicine are accessible to all individuals.

In conclusion, personalized medicine represents a paradigm shift in healthcare, transforming it into a more precise, patient-centered, and efficient model. By leveraging advancements in genomics, technology, and collaboration, personalized medicine has the potential to shape the future of healthcare by improving patient outcomes, enhancing disease prevention efforts, and optimizing healthcare resource utilization. Embracing personalized medicine principles and integrating them into clinical practice will not only revolutionize healthcare delivery but also contribute to the overall well-being and health of individuals and populations.

TAKEAWAY THOUGHT

1. Your DNA Holds the Key: "Genes Unveiled" highlights the incredible power of personalized medicine in utilizing your unique DNA to uncover valuable insights into your health and well-being. By understanding your genetic makeup, you can unlock the secrets encoded in your genes and make informed decisions to optimize your health.

2. Tailored Strategies for Optimal Health: Personalized medicine empowers you with tailored strategies for optimal health and well-being. By considering factors such as genetic predispositions, lifestyle choices, and environmental influences, you can customize interventions that address your individual needs and maximize the potential for positive health outcomes.

3. Collaboration between Science and You: This book emphasizes the collaborative nature of personalized medicine, bringing together the scientific advancements in genomics and your active participation in managing your health. By embracing this partnership, you can actively engage with healthcare professionals, genetic counselors, and other experts to unlock the potential of personalized medicine.

4. Empowering Your Health Decisions: "Genes Unveiled" emphasizes the importance of empowering you to make informed health decisions. By providing you with a deeper understanding of your genetic information, the book enables you to take charge of your health, advocate for personalized care, and actively participate in shared decision-making with your healthcare providers.

5. Preventive Measures for Future Well-being: Personalized medicine highlights the significance of preventive measures based on your unique genetic profile. By identifying potential risks and taking proactive steps to mitigate them, you can safeguard your future well-being and reduce the likelihood of developing certain diseases or conditions.

6. Embracing Personalized Medicine: "Genes Unveiled" encourages you to embrace personalized medicine as a powerful tool in your quest for optimal health and well-being. By recognizing the potential of your genetic information and incorporating personalized approaches into your lifestyle, you can embark on a transformative journey toward a healthier and more fulfilling life.

7. The Future in Your Hands: This book reminds you that the power of personalized medicine lies in your hands. By embracing your genetic information, actively seeking knowledge, and leveraging advancements in personalized medicine, you become an essential participant in shaping the future of healthcare and unlocking the secrets of your unique DNA.

In conclusion, "Genes Unveiled: The Power of Personalized Medicine in Your Hands" empowers you to harness the potential of personalized medicine to optimize your health and well-being. It highlights the importance of collaboration, informed decision-making, and preventive measures, while emphasizing the transformative role you play in unlocking the secrets of your unique DNA for a healthier and more fulfilling life.

Made in the USA
Las Vegas, NV
24 September 2023